The Antietam Battlefield Atlas

Brad Butkovich

Copyright © 2023 Historic Imagination LLC

www.historicimagination.com

All rights reserved.

ISBN 978-1-7325976-5-5

Table of Contents

Introduction v

Map Legend vi

Prelude to Battle 1

The First Corps Begins the Battle 7

The Twelfth Corps Attacks 25

The West Woods 37

The Sunken Road 69

The Middle Bridge 99

Burnside's Bridge 109

Ninth Corp Advance and Hill's Counterattack 119

Order of Battle 139

Bibliography 153

Notes 157

Introduction

The Battle of Antietam was the single bloodiest day in the American Civil War. Of course, several other battles had higher casualty counts: Gettysburg, Chickamauga, and The Wilderness to name a few. However, those battles spanned several days. At Antietam, the approximately 12,000 Union and 10,000 Confederate killed wounded, and missing all occurred in one day. In fact, the majority in just a few hours, so intense and devastating was the struggle.

Attempting to sort out the ebb and flow of a battle with so many men, often engaged in such a relatively small area, can be a challenge for historians. But by using a variety of primary sources, we tend to give it our best shot. The father of Antietam historiography was Ezra A. Carman. He fought at the battle as colonel of the 13th New Jersey Infantry. After the war, he was appointed to several boards for the battle, including for the National Cemetery and park. He spent much of his later life writing a manuscript on the fight along Antietam Creek, and was instrumental in publishing a series of fourteen highly detailed maps of the battlefield. Later historians have used these maps as a base, and by consulting new research and discovering new primary sources, a reasonably accurate interpretation of the battle can be achieved.

This atlas is the product of years of research that originally began when I started writing the *Visual Antietam* series. *Visual Antietam* is a printing of the Carman manuscript with a heavy emphasis on maps and pictures to complement and enhance the text. Now that the series has been out for several years I have decided to update the maps and compile them together into one paperback volume. I feel this format allows the reader to easily carry it while walking the battlefield. And unlike the *Visual Antietam* series, this atlas is in full color.

The book is laid out with one full color map per page. Highlights of the action shown on the map are listed underneath. A rough time estimate is given for each map. However, historic times are an estimate at best. The participants didn't write down times in the middle of the battle, and when officers did, they didn't have their 19th Century spring-wound watches synchronized to each other. The times given are a best guess based upon sources such as the Carman maps and officer reports, and should be viewed more as a progression of the action than an ironclad time stamp.

Readers interested in a more in-depth study of the battle should read either the *Visual Antietam* series, or Tom Clemens' reprint of the Carman manuscript which is fully annotated and footnoted with a breakdown and analysis of the primary sources involved: *The Maryland Campaign of September 1862: Vol. II: Antietam*. Clemens' book is used extensively in the endnotes as it leads to mentions of extensive correspondence between Carman and veterans, as well as records in the National Archives. Carman's original maps are also easily accessible from the Library of Congress website, and are very much worth the download.

I hope you enjoy the 124 full color maps, and find them useful while walking the field and discovering the battle. If you had an ancestor present at Antietam, I hope this atlas helps you find where they fought, so that you can stand in the same spot. Happy battlefield stomping!

Map Legend

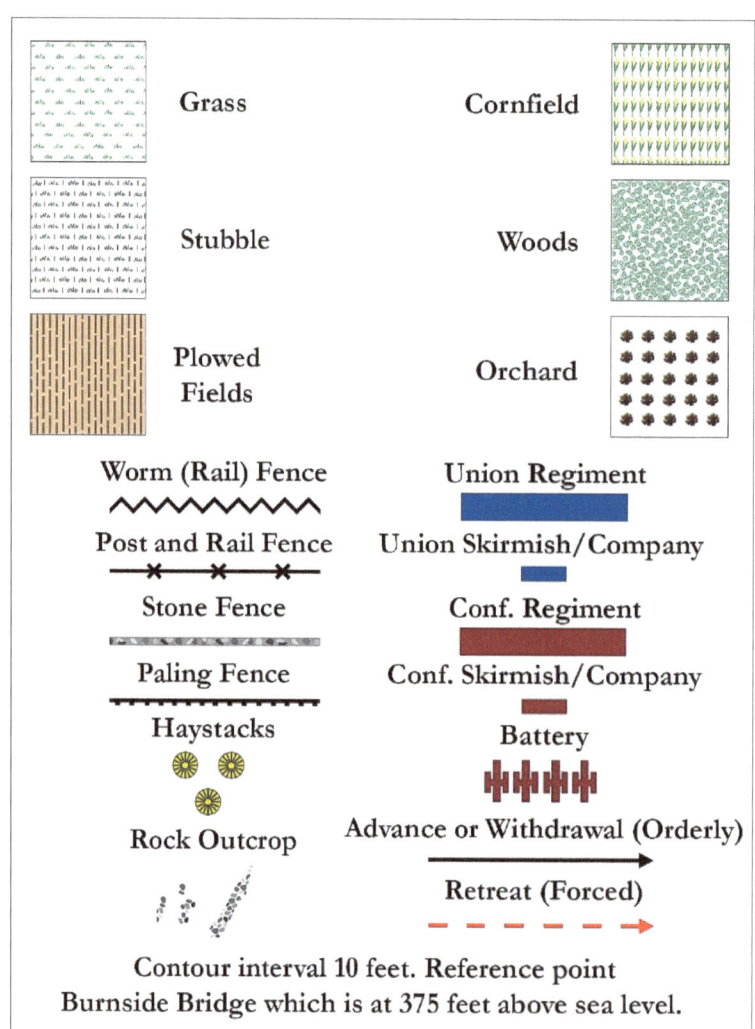

Prelude to Battle
September 16th, 1862

After the Confederate victory at Second Bull Run the leaders of the rebellion decided to invade the North. General Robert E. Lee, commanding the Army of Northern Virginia, intended to bring the battle to northern territory. President Jefferson Davis believed that a southern victory on Union soil could lead to foreign recognition of the Confederacy. Both believed that the slave state of Maryland would welcome the southern army and bring much needed recruits. Logistics was also a factor. The farms of northern Virginia had been stripped bare of food and resources over the last year. The farms north of the Potomac River would help feed Lee's men.

Lee's army crossed the Potomac into Maryland on September 3rd. The Union Army of the Potomac, then stationed around Washington D.C. to protect the capital after their defeat, marched northwest to intercept them. A portion of Lee's army captured the Union garrison at Harper's Ferry, while the rest continued north. Two Union soldiers captured a copy of Lee's battle plan and presented them to the Union commander, Major General George B. McClellan. McClellan moved to intercept Lee's divided army. However, a stubborn defense of the passes through South Mountain on September 14th delayed the federals. By the 15th McClellan's army was at the banks of Antietam Creek, normally crossed by three stone bridges appropriately named the Upper, Middle, and Lower Bridges. The Army of Northern Virginia was on the heights beyond waiting for them. Lee had decided to stay and give battle. He only needed time to reunite his divided forces.

McClellan gave him that time. Instead of attacking vigorously the next day, September 16th, he waited. No Union cavalry was tasked to envelope and ascertain the Confederates lines during the day. Nor were any infantry for that matter. Only in the afternoon did McClellan order Major General Joseph Hooker and his First Corps to cross the creek and probe the north end of the enemy position. He then ordered his Twelfth Corps to cross over during the night, to be supported by the Second Corps the next morning, and the Sixth Corp if necessary. The Ninth Corps would remain at the left flank near the Lower Bridge.

Hooker crossed the Antietam about 4 p.m. at the Upper Bridge and nearby fords. The 3rd Pennsylvania Cavalry led the advance, followed by Brigadier General George G. Meade's division of Pennsylvania Reserves. The cavalry skirmished with the Confederates at the East Woods, before Brigadier General Truman Seymour's brigade of Reserves charged into the woods at dusk and cleared it. Brigadier General John B. Hood's division moved north into the East Woods to support the Confederate skirmish line, and night fell with the two sides in close proximity.

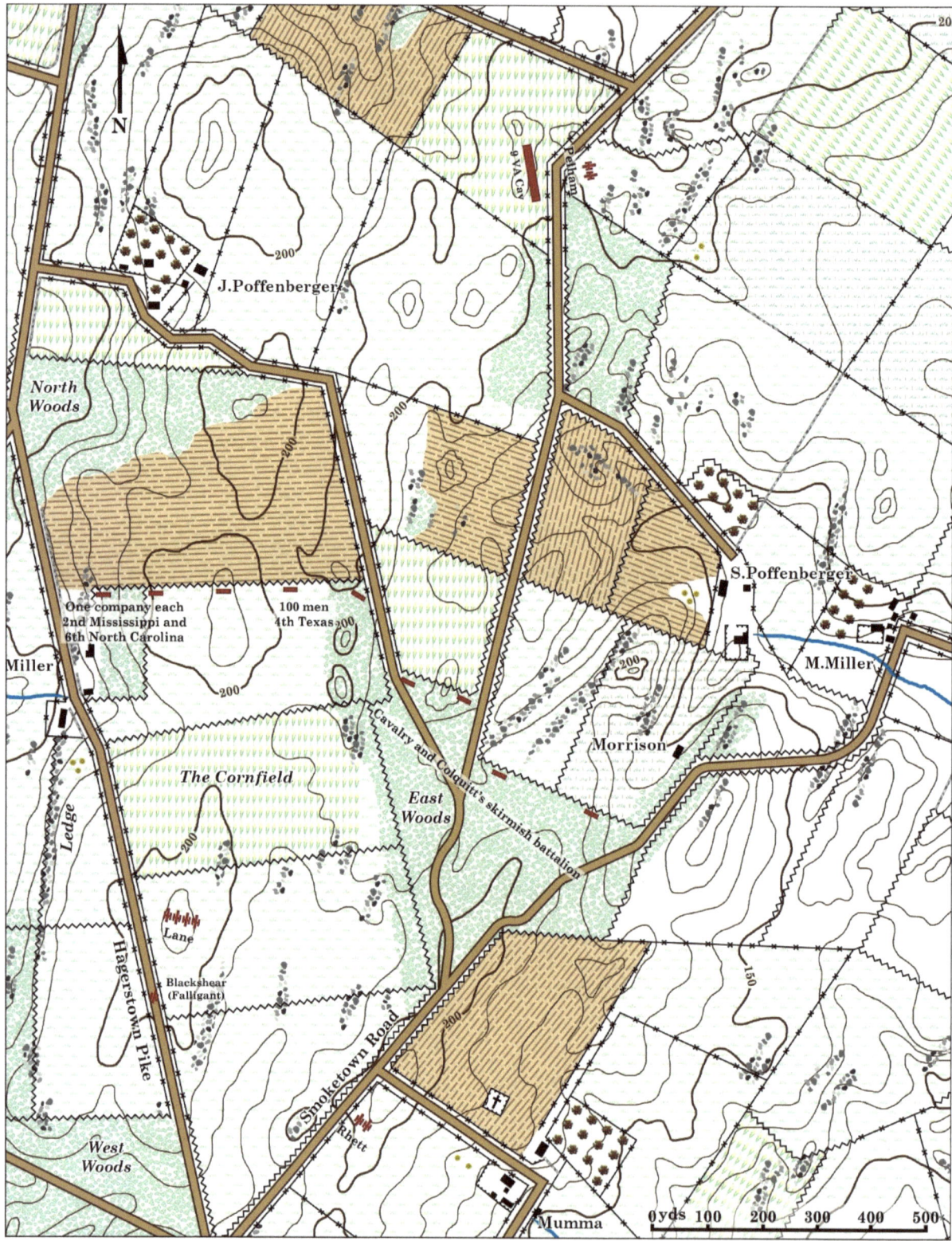

1. Afternoon.
2. Initial deployment of the Confederates during the late afternoon and early evening of September 16th.
3. The balance of the Confederate forces are in the West Woods.

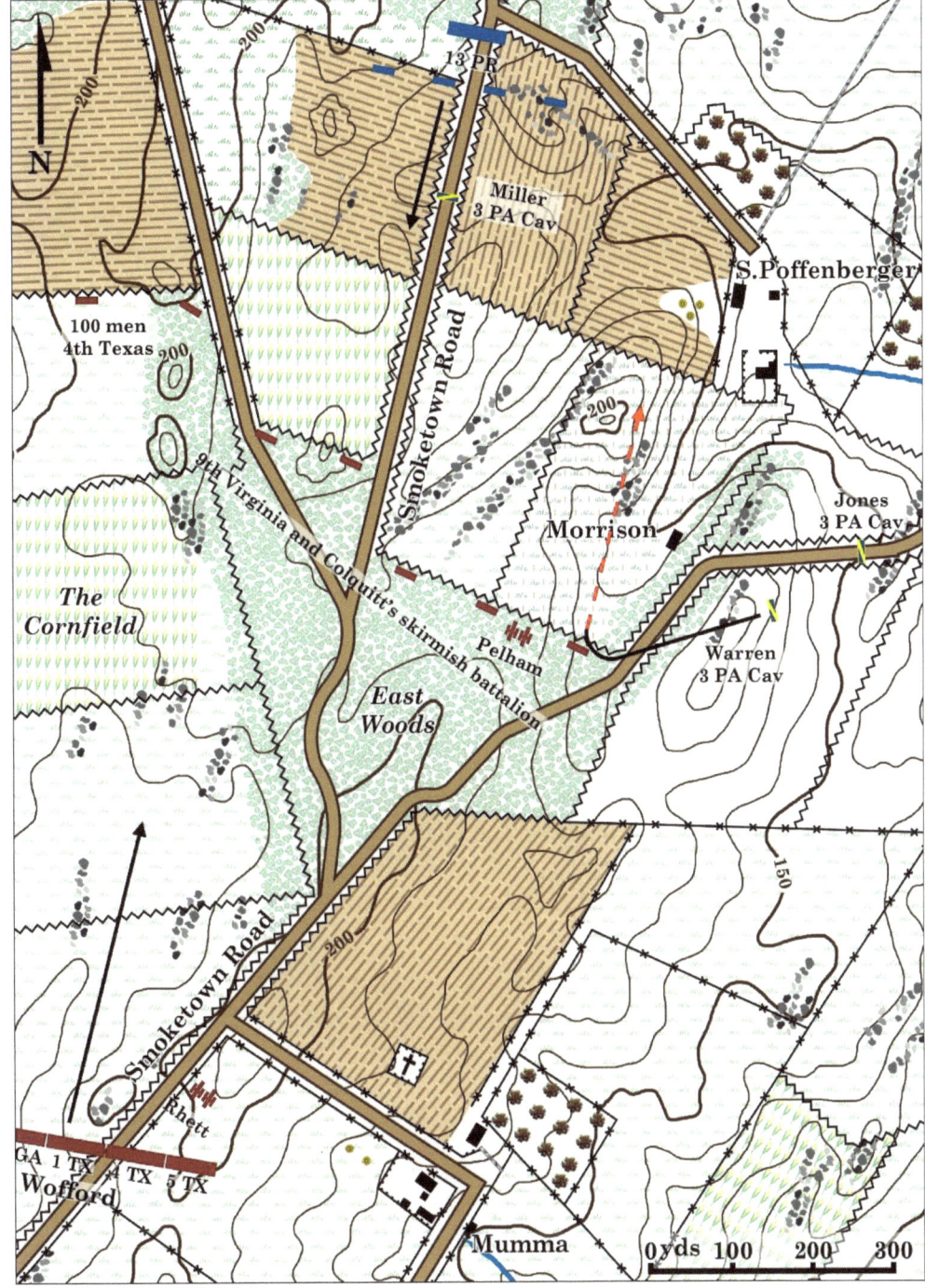

1. Twilight. Approximately 6:15 p.m.
2. Captain Edward S. Jones' squadron of the 3rd Pennsylvania Cavalry approaches the East Woods from the east from the M. Miller farm. Lieutenant E. Willard Warren's platoon of Company C in Jones' squadron charges the woods. They encounter Confederate skirmishers and Pelham's battery. Despite rifle and canister fire, they take no casualties and turn north back into Union lines.[1]
3. Lieutenant William E. Miller's Company H, 3rd Pennsylvania Cavalry covers the approach on the Smoketown Road. They fall back from Confederate rifle fire into the lines of the advancing 13th Pennsylvania Reserves "Bucktails".[2]
4. Wofford's Texas Brigade moves north to support the skirmishers. Pelham will fall back.

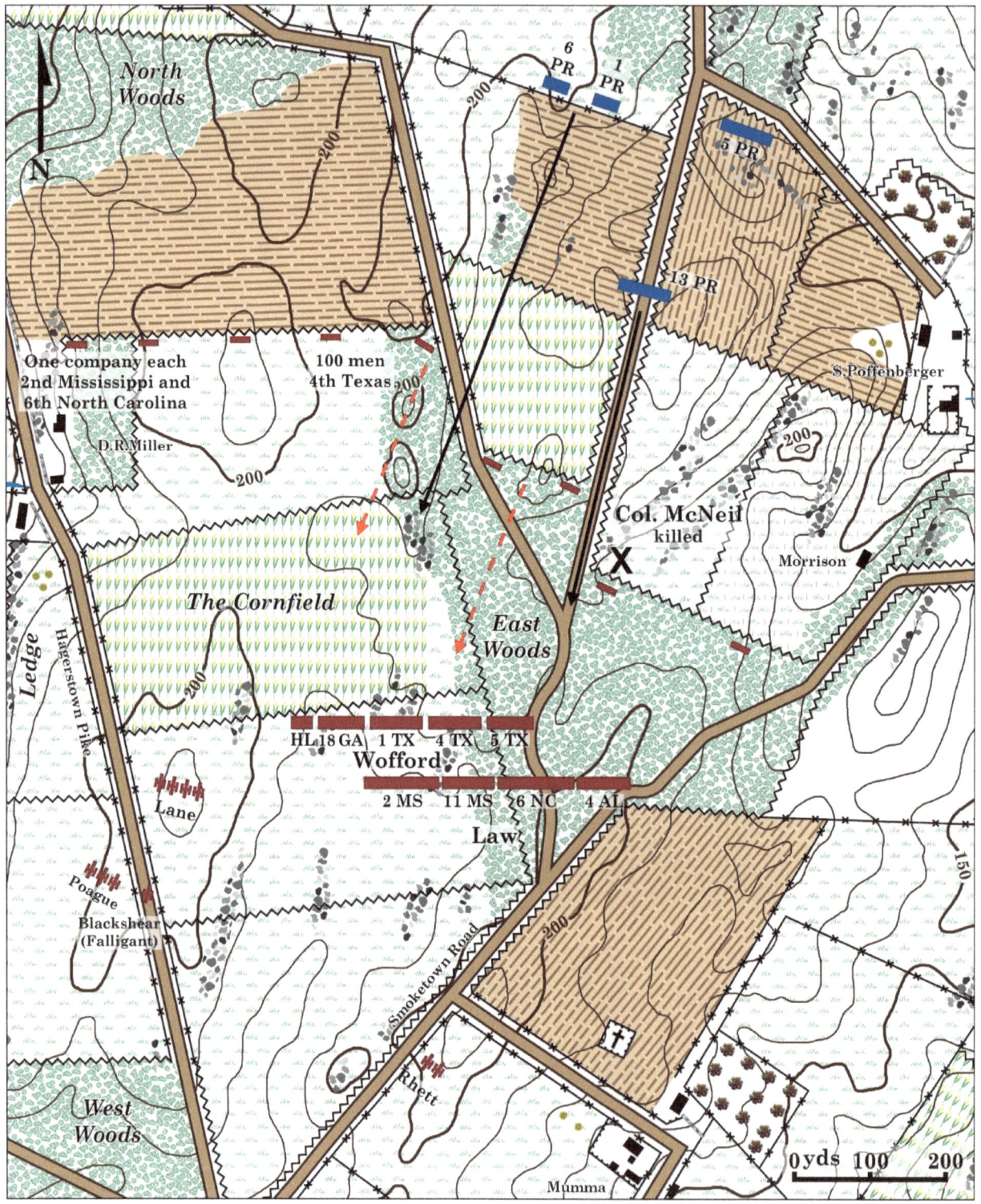

1. 6:30 p.m. to 10:00 p.m.
2. John B. Hood's division has moved forward to support the advance skirmish lines.
3. Truman Seymour's brigade moves into the East Woods, forcing the skirmishers there to fall back.
4. Colonel Hugh W. McNeil commanding the 13th Pennsylvania "Bucktails" leads the advance. He is killed on the left side of the Smoketown Road fifteen yards from the tree line.[3]

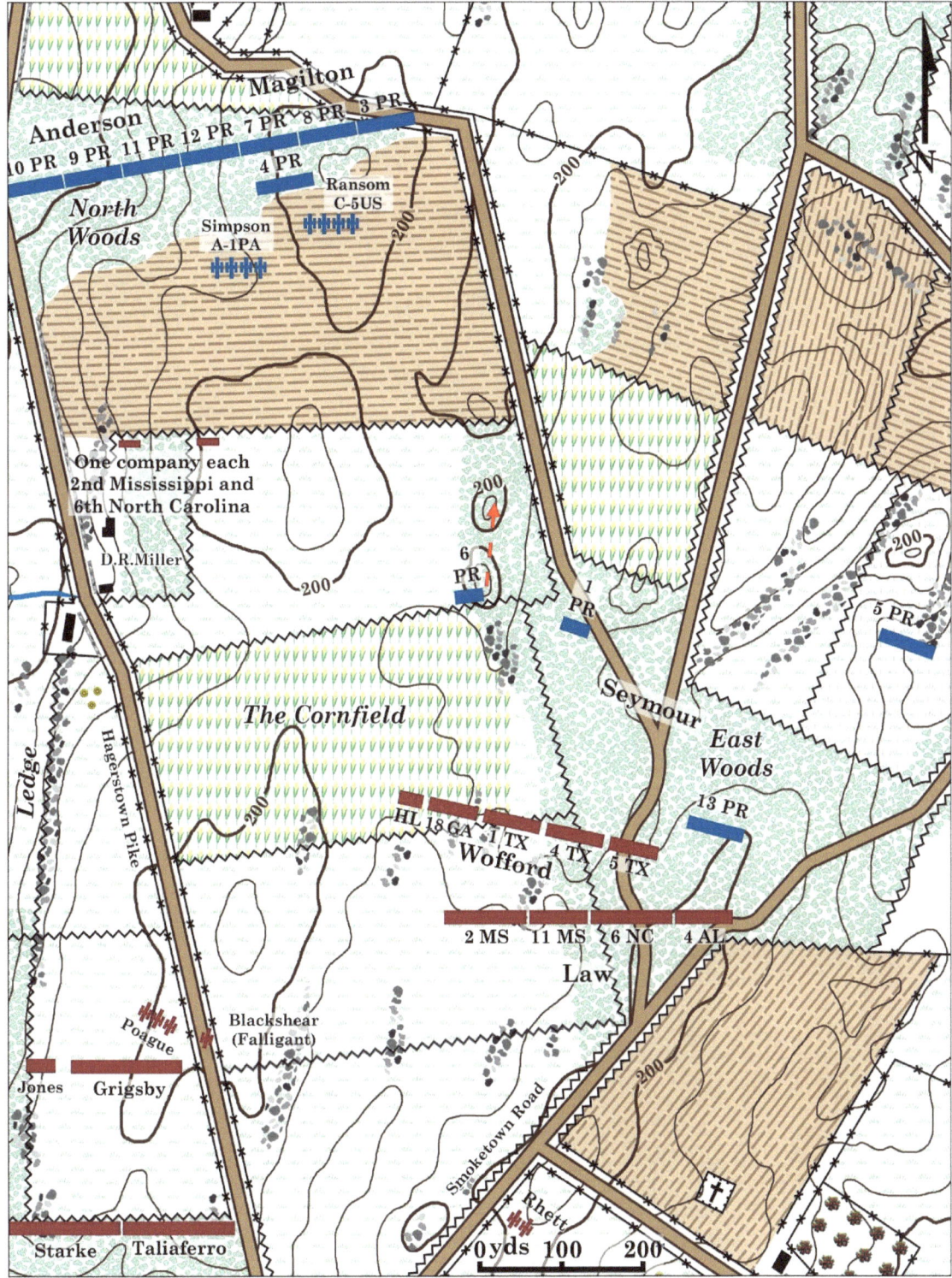

1. Night.
2. The 6th Pennsylvania Reserves enter the Cornfield, but fire from Wofford pushes them back.[4]
3. The 13th Pennsylvania Reserves end the day in close proximity to Hood's Division.[5]
4. Hood will receive orders to pull back during the night.[6]
5. Jackson's Division, now commanded by John R. Jones, takes up a position west of the Hagerstown Turnpike

The First Corps Begins the Battle

The Battle of Antietam began at first light, as soon as the soldiers could see each other. Seymour's brigade advanced through the East Woods to find that Hood had withdrawn during the night. The Confederates now had a solid line facing north. Lieutenant General Thomas J. "Stonewall" Jackson's command held the north end of the Confederate line around Sharpsburg. Lieutenant General James Longstreet commanded the center and south. Most of Jackson's command was near the West Woods. Ewell's Division, commanded by Brigadier General Alexander R. Lawton, had two brigades to the east of the Hagerstown Turnpike, and two west. Jackson's Division, commanded by Brigadier General John R. Jones, lay west of the pike. Hood's Division was in reserve in the West Woods. A brigade from Major General Daniel H. Hill's division supported Lawton, and Colonel Stephen D. Lee's Battalion of artillery supported them all from a plateau east of the Dunker Church. Finally, several batteries of Confederate horse artillery dominated the western area from their position on Nicodemus Hill.

Seymour's brigade immediately engaged Lawton. Brigadier General James B. Rickett's division moved south to support him. General Hooker sent Brigadier General Abner Doubleday's division south along the pike, but they were held up by Confederate skirmishers. Rickett's division arrived piecemeal. Brigadier General George L. Hartsuff was wounded, and Colonel William A. Christian fled the field, delaying the advance of both brigades. Brigadier General Abram Duryée's brigade attacked alone through the David R. Miller Cornfield. His brigade was repulsed, and Hartsuff and Christian's brigades moved through them. They were met with the same fire from Lawton and stopped. Doubleday's division advanced, led by Brigadier General John Gibbon's Iron Brigade. Together they stopped an advance by Lawton, and Doubleday moved closer to the church. An attack by Jones' Division was repulsed after a savage close range firefight along the turnpike fences.

Hood's Division then filed out of the West Woods and charged into the oncoming federals in a devastating counterattack. They chased the Union back across the cornfield to its northern fence, and cleared the East Woods. The initiative had temporarily swung in favor of the Confederates.

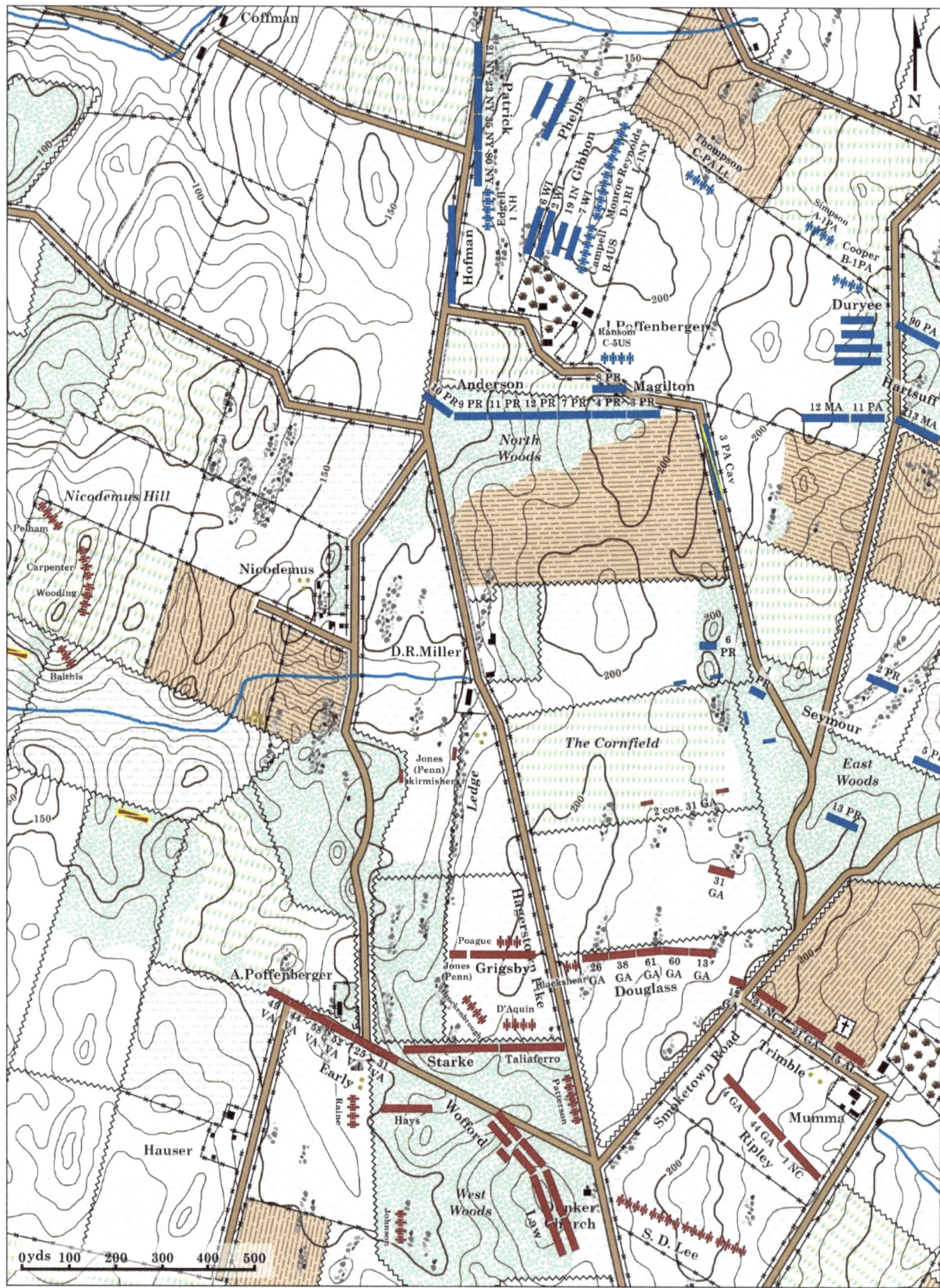

1. The opposing forces at first light.

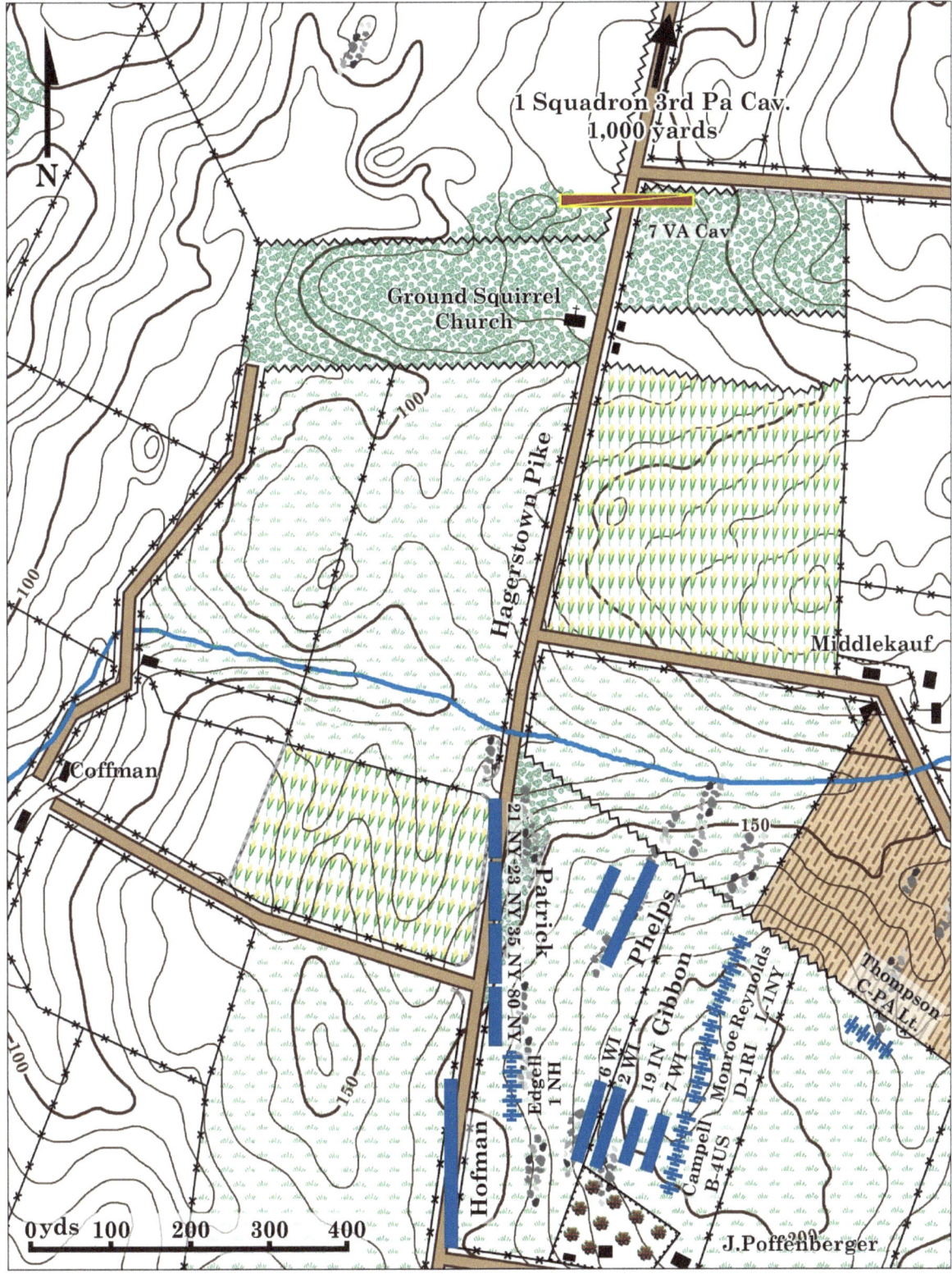

1. Dawn.
2. The 7th Virgina Cavalry is stationed at Ground Squirrel Church to the north and cut off from the army's main body. The Union First Corps is to the south and a squadron of the 3rd Pennsylvania Cavalry is located along the Hagerstown Turnpike approximately 1,000 yards to the north.[1]

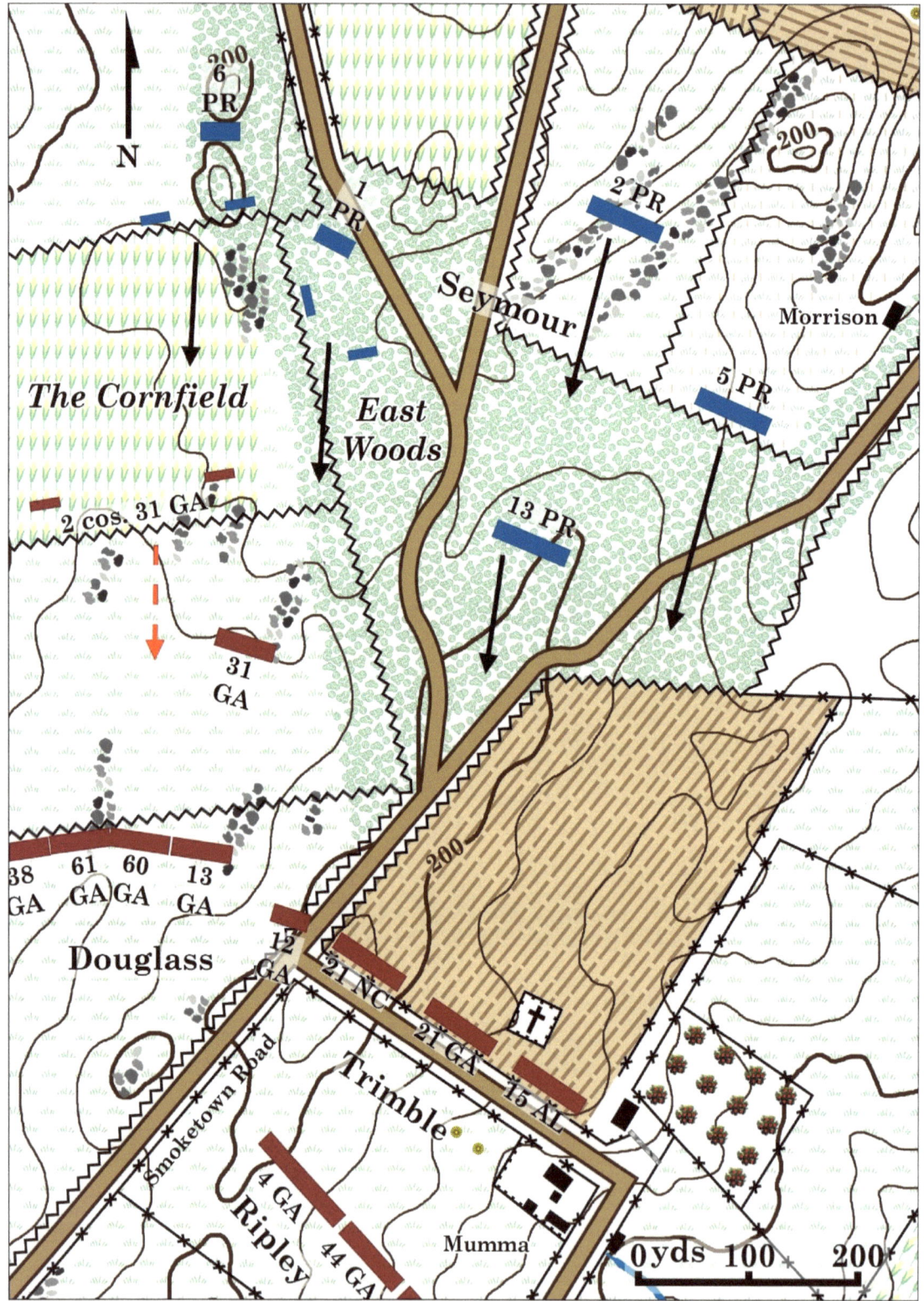

1. First light. Approximate 5-5:15 a.m.
2. Seymour's brigade advances.
3. Skirmishers of the 1st and 6th Pennsylvania Reserves push two companies of the 31st Georgia out of the cornfield.[2]

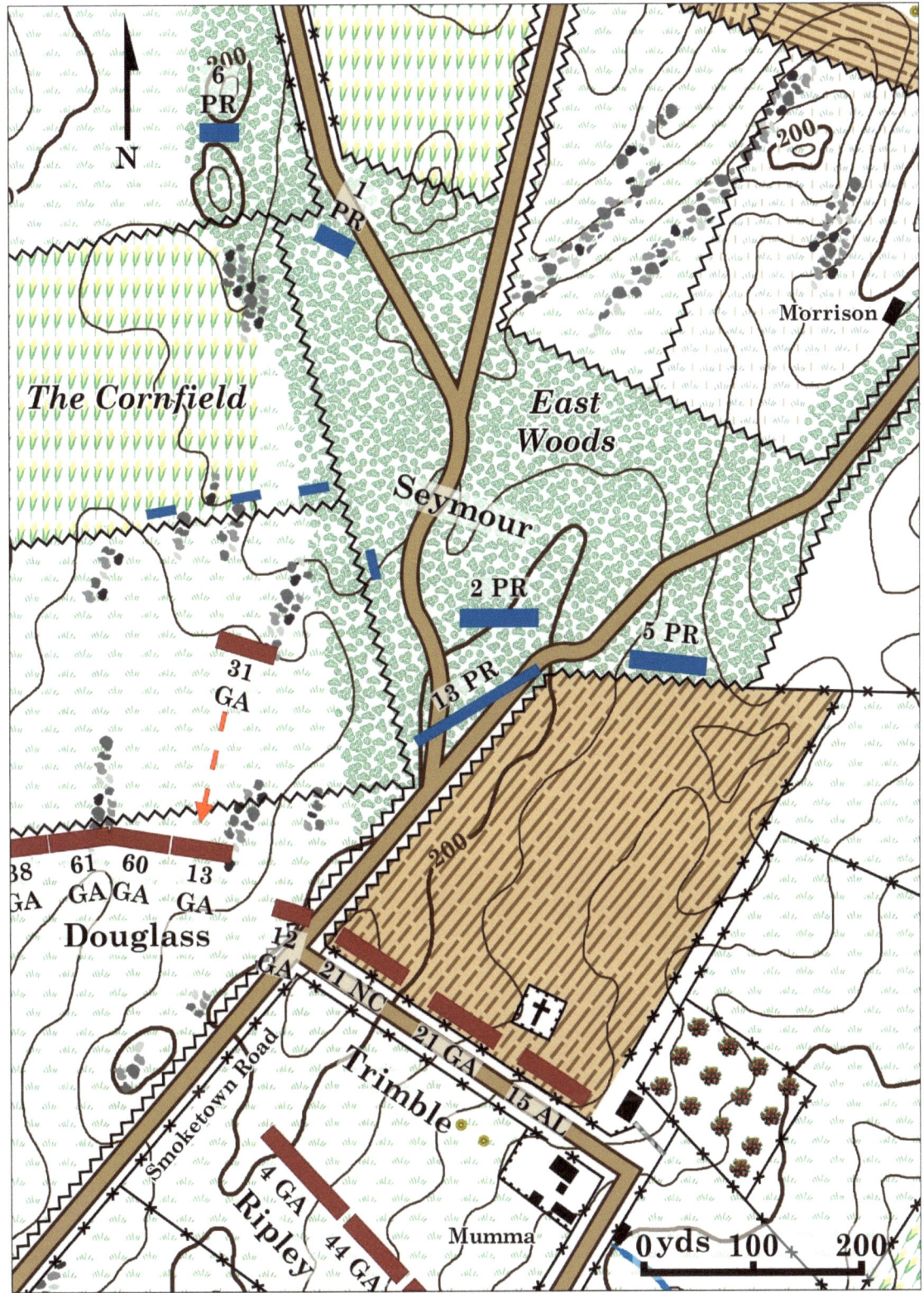

1. 6:00 a.m. to 6:20 a.m.
2. Fire from the 1st and 6th Pennsylvania Reserve skirmishers force the 31st Georgia to retire back to Douglass' main line.[3]
3. Trimble's Brigade, commanded by Colonel James A. Walker, engages Seymour at the wood line.

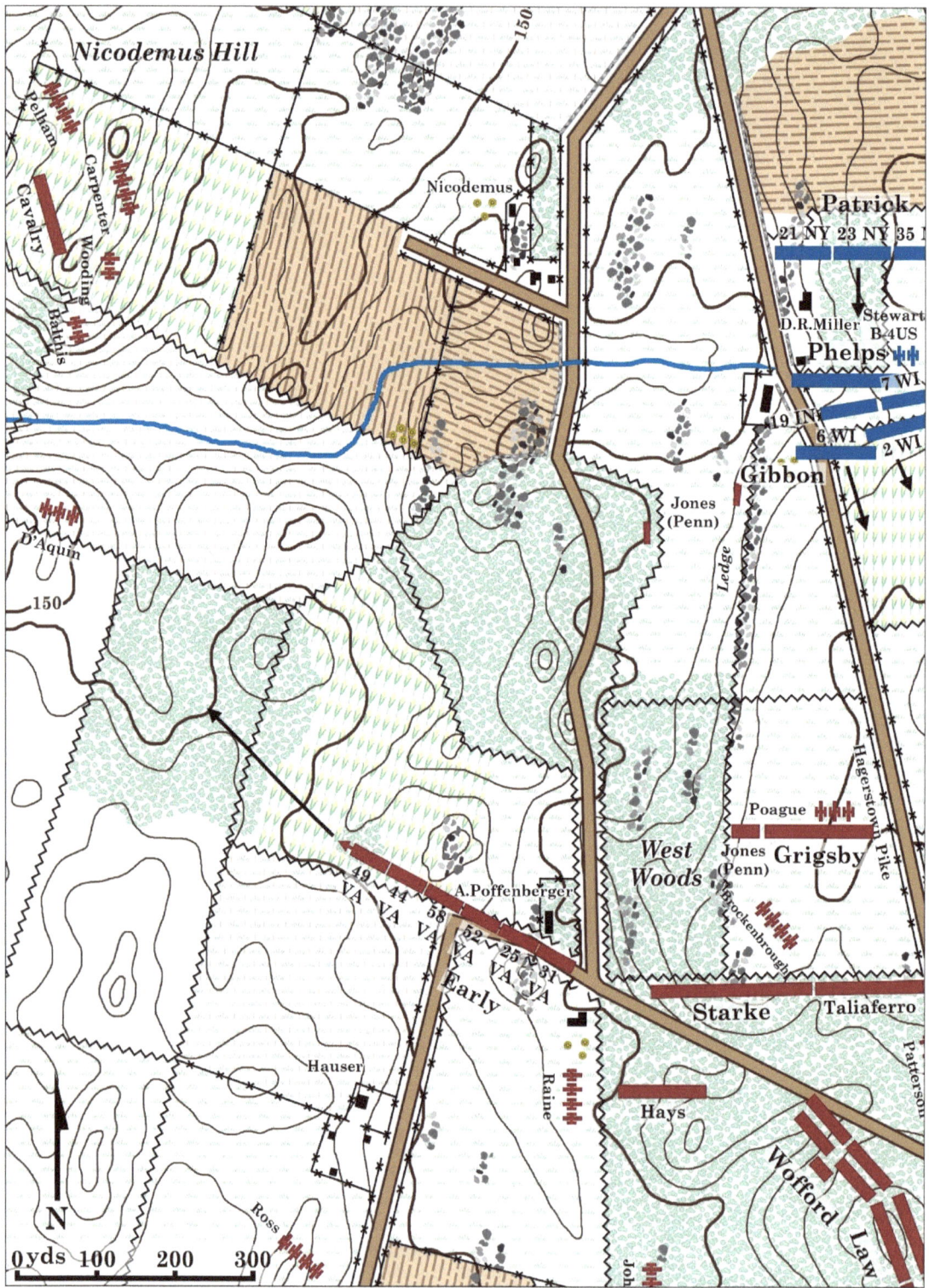

1. 6:10 a.m.
2. As Doubleday's division advances along the Hagerstown Turnpike, Early's Brigade is sent toward Nicodemus Hill to help provide support.
3. The Confederate batteries on Nicodemus Hill will fire into the flank of the Union infantry around the Miller farm, and engage the Union artillery at the J. Poffenberger farm all day.

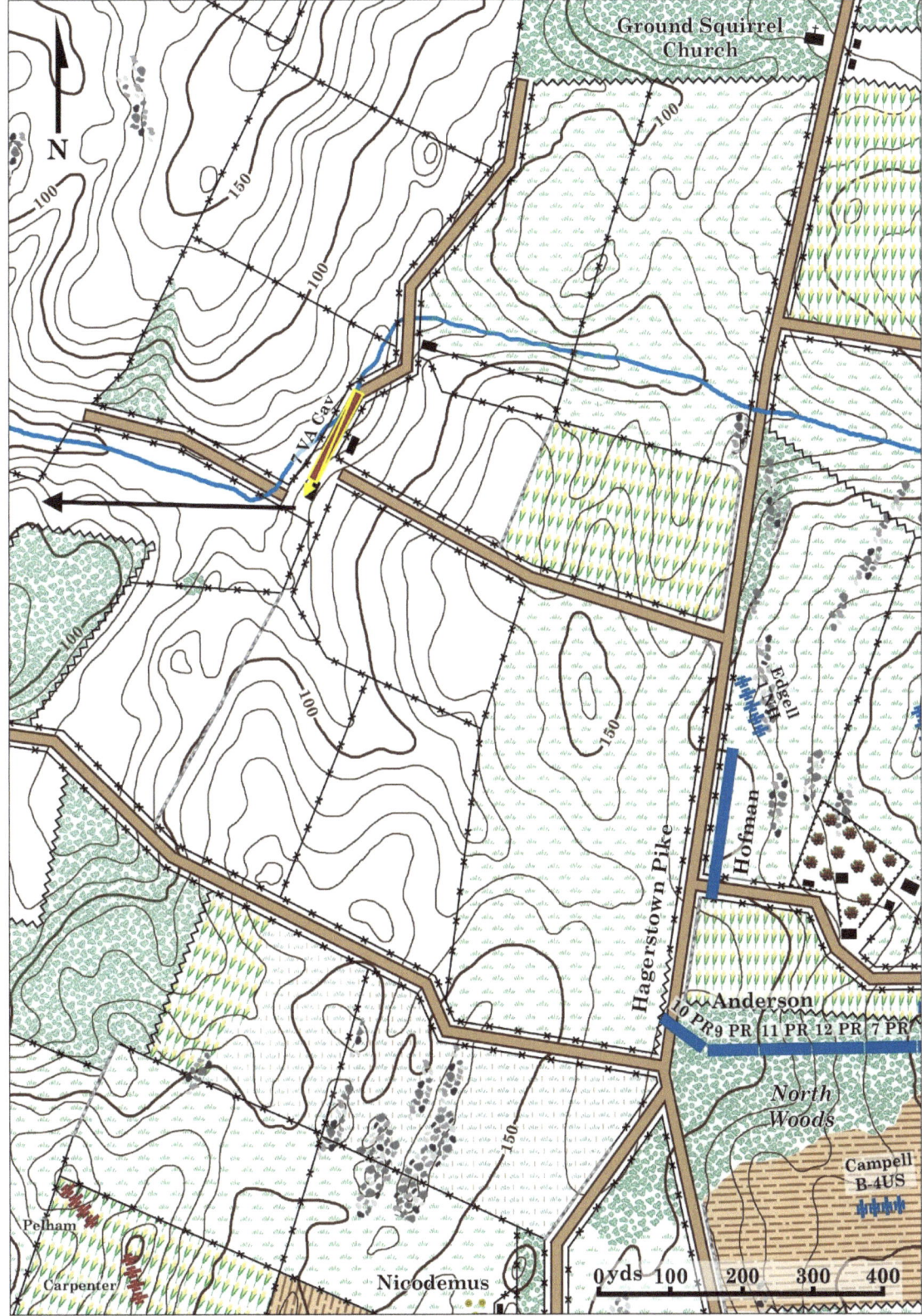

1. 6:20 a.m.
2. The 7th Virginia Cavalry moves southwest away from Ground Squirrel Church to New Industry on the Potomac River. It will move south and rejoin the army at Nicodemous Hill around 9 a.m.[4]

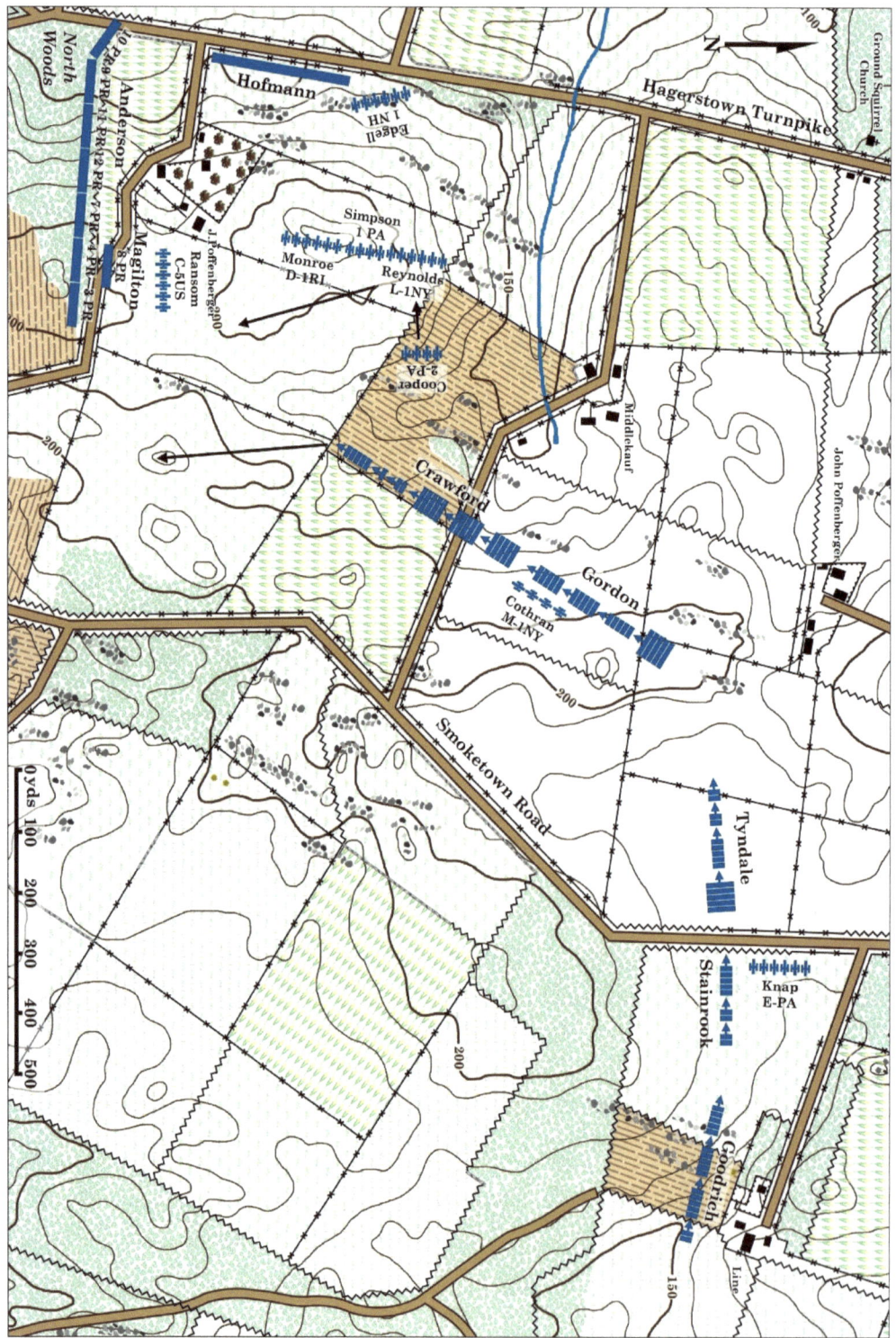

1. 6:30 a.m.
2. The Twelfth Corps marches to the battlefield from its camps northeast of the Miller farm.
3. Corps commander Major General Joseph K. Mansfield orders his regiments to approach in narrow column by divisions, which is two companies wide and four to five deep, depending on the number of companies present. He feels it will be easier for many of his untried regiments to maneuver to the battlefield in this formation before forming into line of battle closer to the enemy.[5]

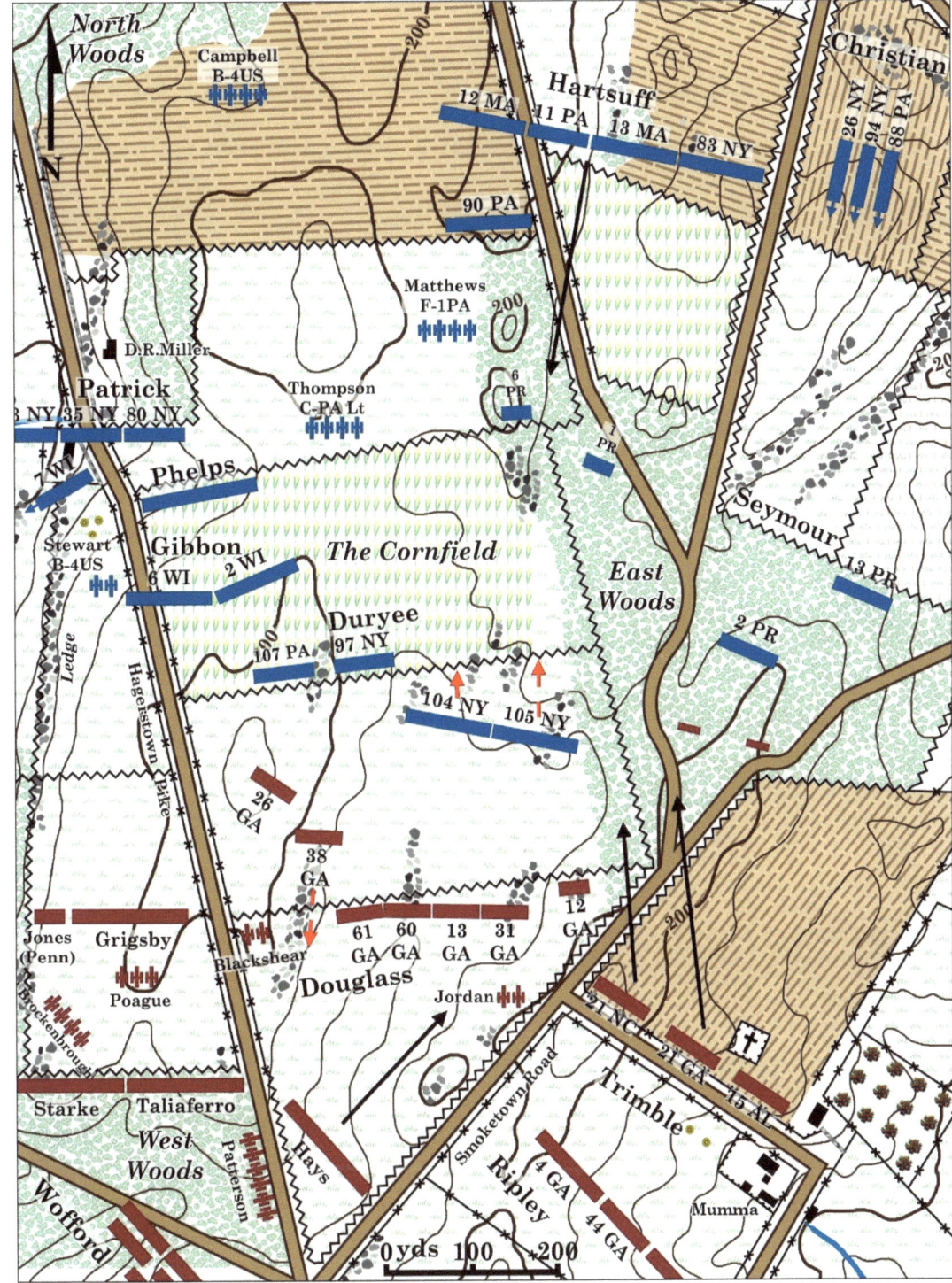

1. 6:30 a.m.
2. Duryée's brigade advances and engages Douglass. The 104th and 105th New York move into the field south of the Cornfield, but are forced to retreat by the heavy fire.[6]
3. The 26th Georgia moves to high ground to enfilade Duryée. The 38th Georgia advances to a rock ledge for better cover, but fire forces them back.[7]
4. Seymour pulls back from the East Woods and Walker (Trimble) advances.
5. Hay's Brigade moves forward to support Douglass.

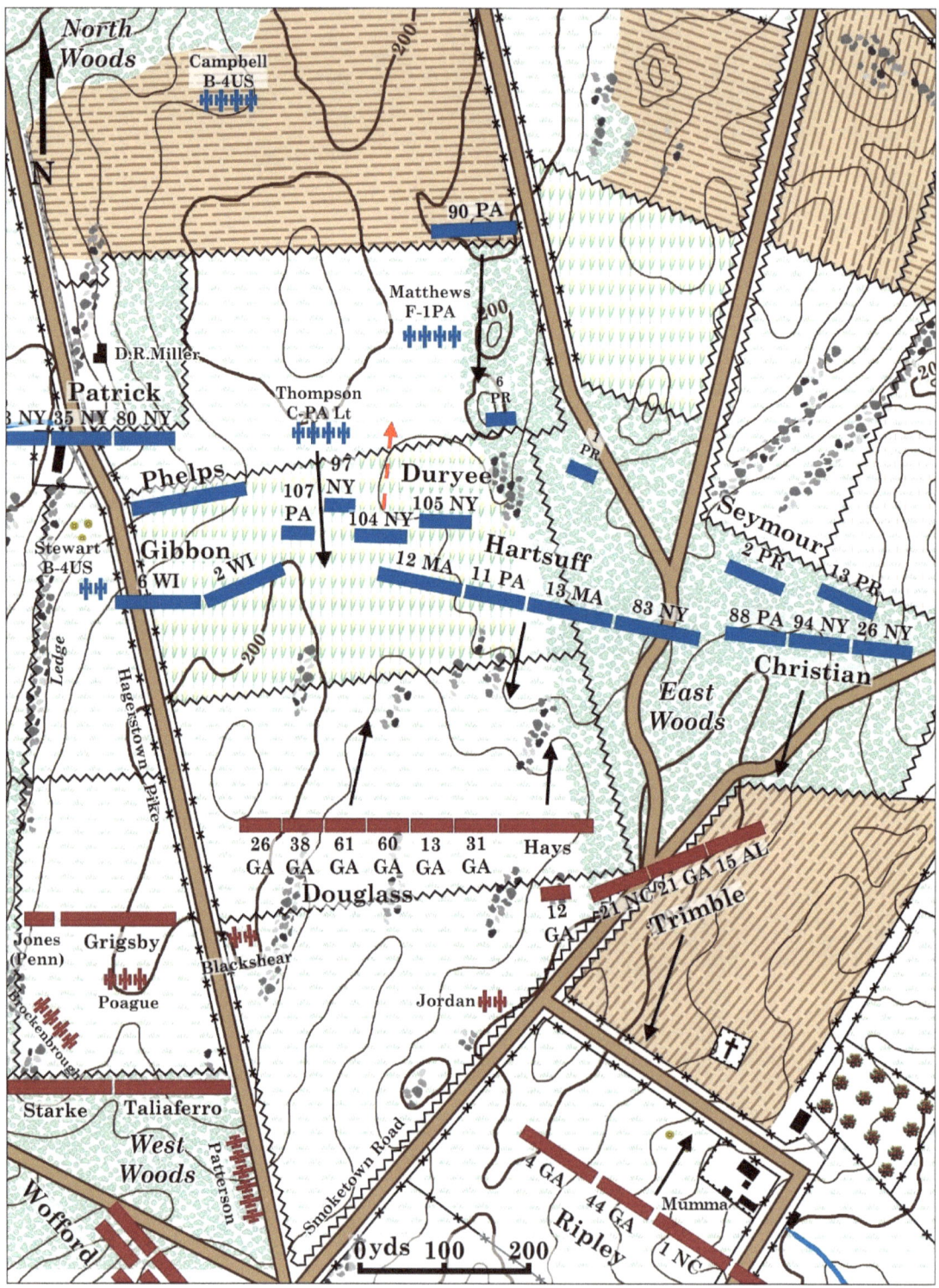

1. 6:40 a.m.
2. Duryée's brigade falls back.
3. Hartsuff advances through the Cornfield and West Woods. Christian deploys and advances to his left.
4. Walker pulls back Trimble's Brigade after realizing Douglass and Hays have not advanced as far and they are exposed.[8]
5. Douglass and Hays advance and pursue Duryée.

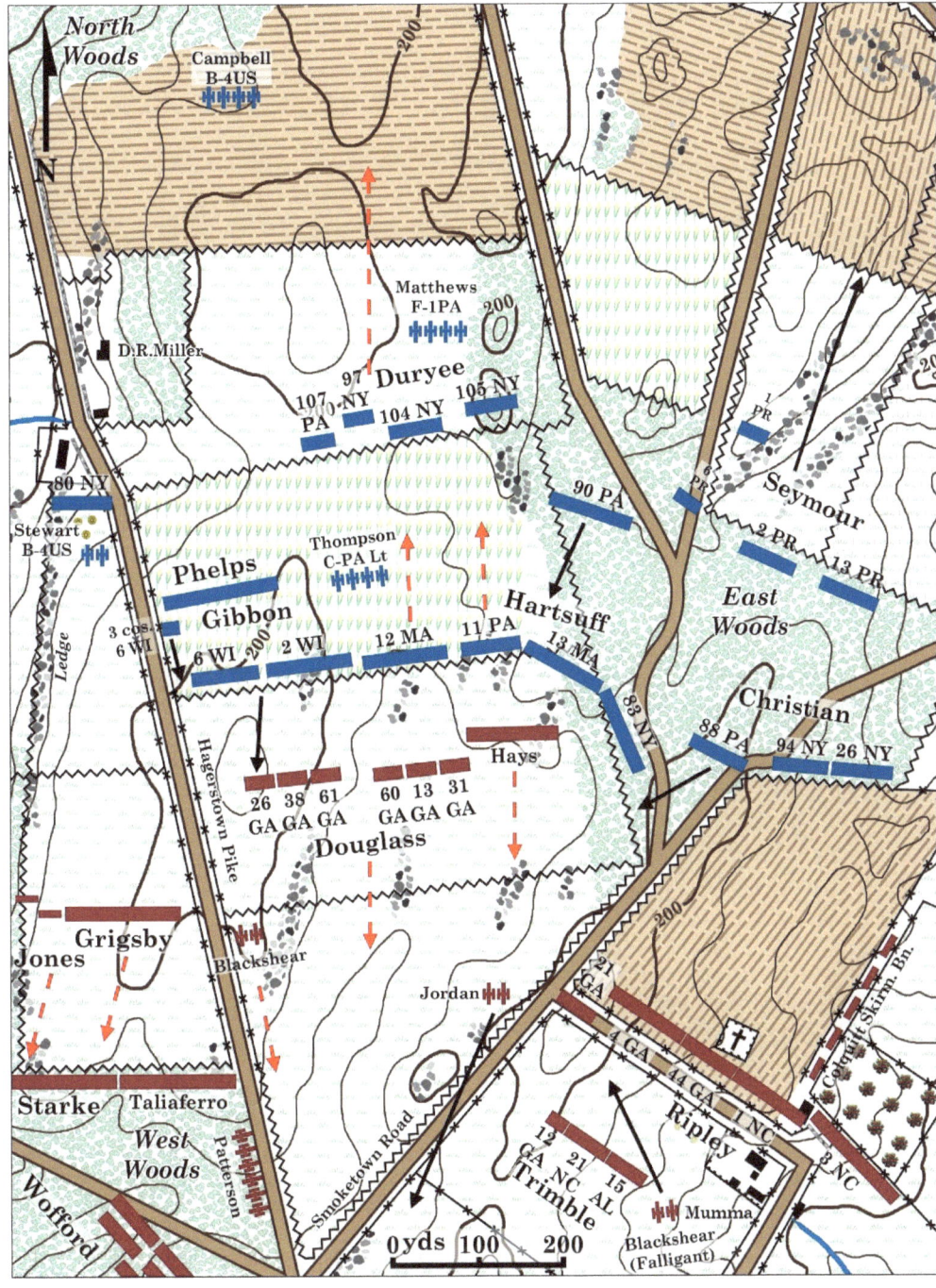

1. 6:50 a.m.
2. Ripley's Brigade passes through Walker (Trimble). General Ripley is wounded and command passes to Colonel George P. Doles of the 4th Georgia.[9]
3. Douglass and Hays devastate Hartsuff's right wing. The 90th Pennsylvania marches through Hartsuff and engages the Confederates. The 12th Massachusetts and 11th Pennsylvania fall back after the 90th passes through them.[10]
4. Gibbon arrives near the Hagerstown Pike, and Hartsuff's left wing and the 88th Pennsylvania swing to the west. Caught in the flank, Douglass and Hays fall back.[11]
5. The left wing of Gibbon's Iron Brigade follows Douglass, forcing Jones' and Grigsby's Brigades to fall back to the West Woods.[12]

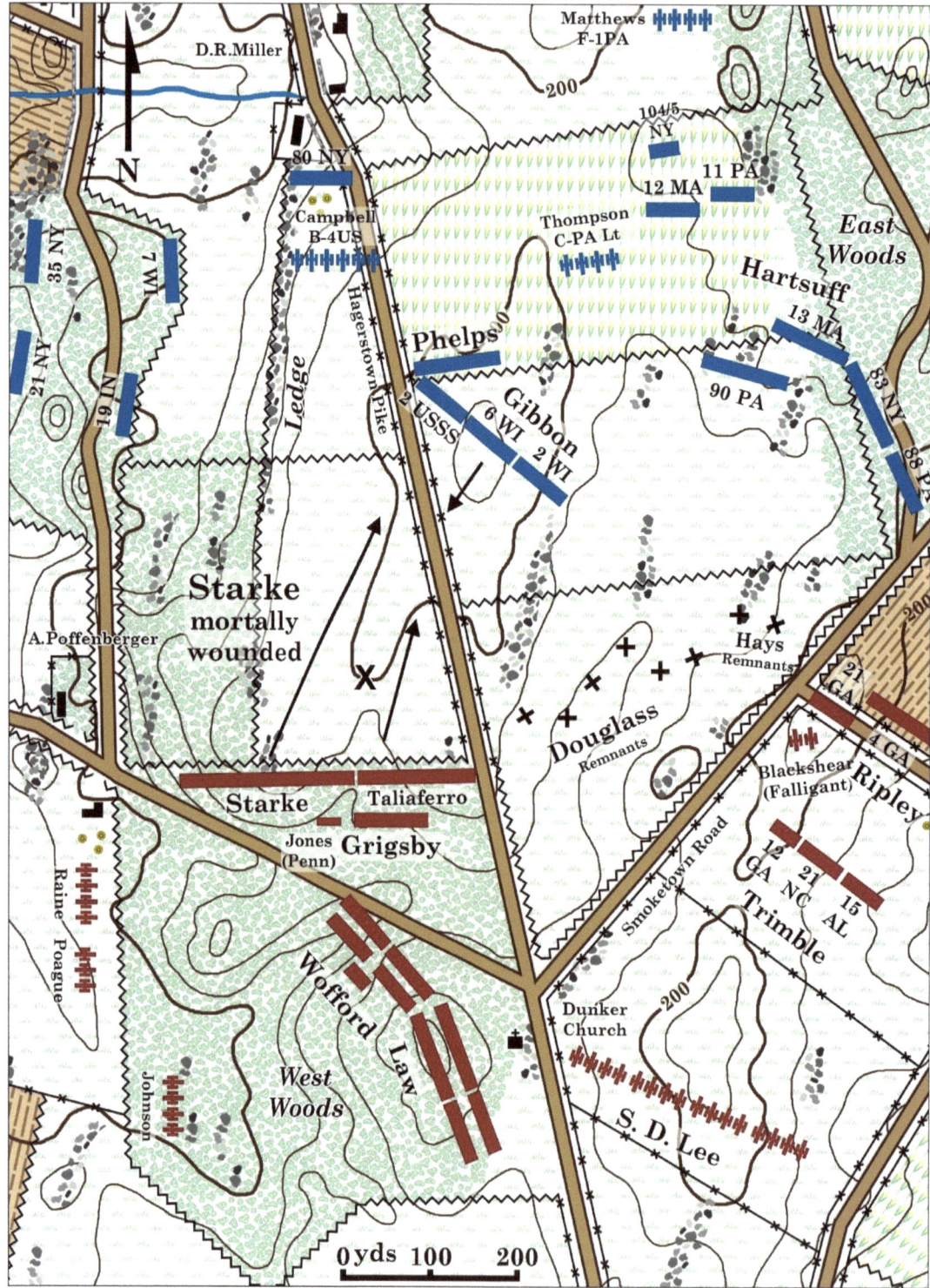

1. 6:55 a.m.
2. Starke and Taliaferro's Brigades march out of the West Woods and turn to the right to confront Gibbon and Phelps.
3. Brigadier General William E. Starke is mortally wounded moving his brigade forward.[13]
4. The Union brigades turn west and the two sides fight between the stout rails of the Hagerstown Turnpike fence.[14]

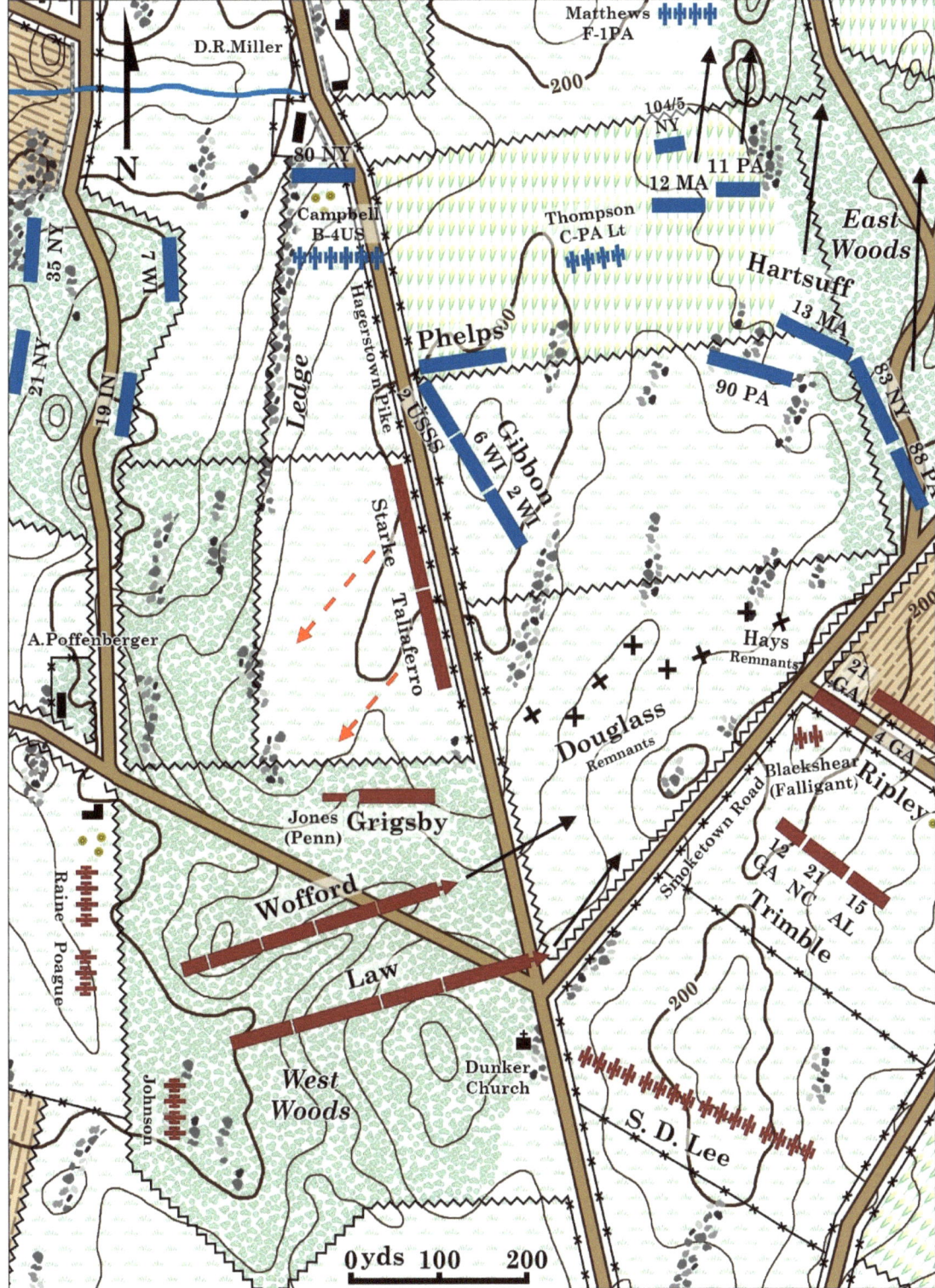

1. 7:05 a.m.
2. The pressure is too great, and Starke and Taliaferro's brigades retreat to the West Woods. As the Union follows up and moves south down the turnpike, Hood orders his division forward to stop them. They move into the fields south of the Cornfield and deploy.[15]
4. Christian and Hartsuff's brigades retire, except the 90th Pennsylvania.[16]

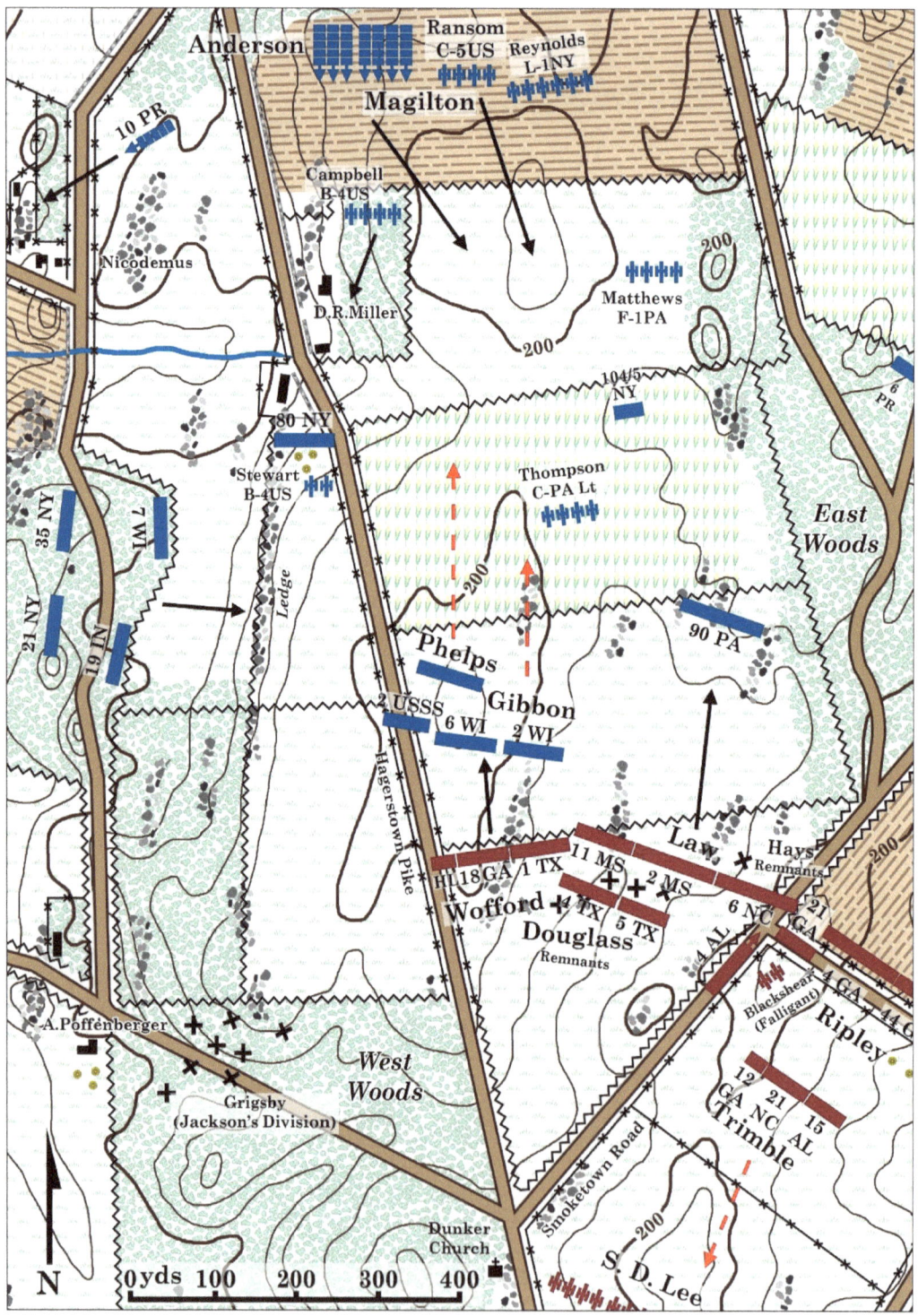

1. 7:10 a.m.
2. Hood charges the oncoming Federals and forces Gibbon's left wing and Phelps to retreat.
3. Gibbon's right wing, the 7th Wisconsin and 19th Indiana, leave the West Woods and advance to the limestone ledge.[17]
4. Magilton and Anderson's brigades from Meade's Pennsylvania Reserve division leave the North Woods and approach the Cornfield.
5. Trimble's brigade retires to the south. The 21st Georgia remains at the Mumma lane.[18]

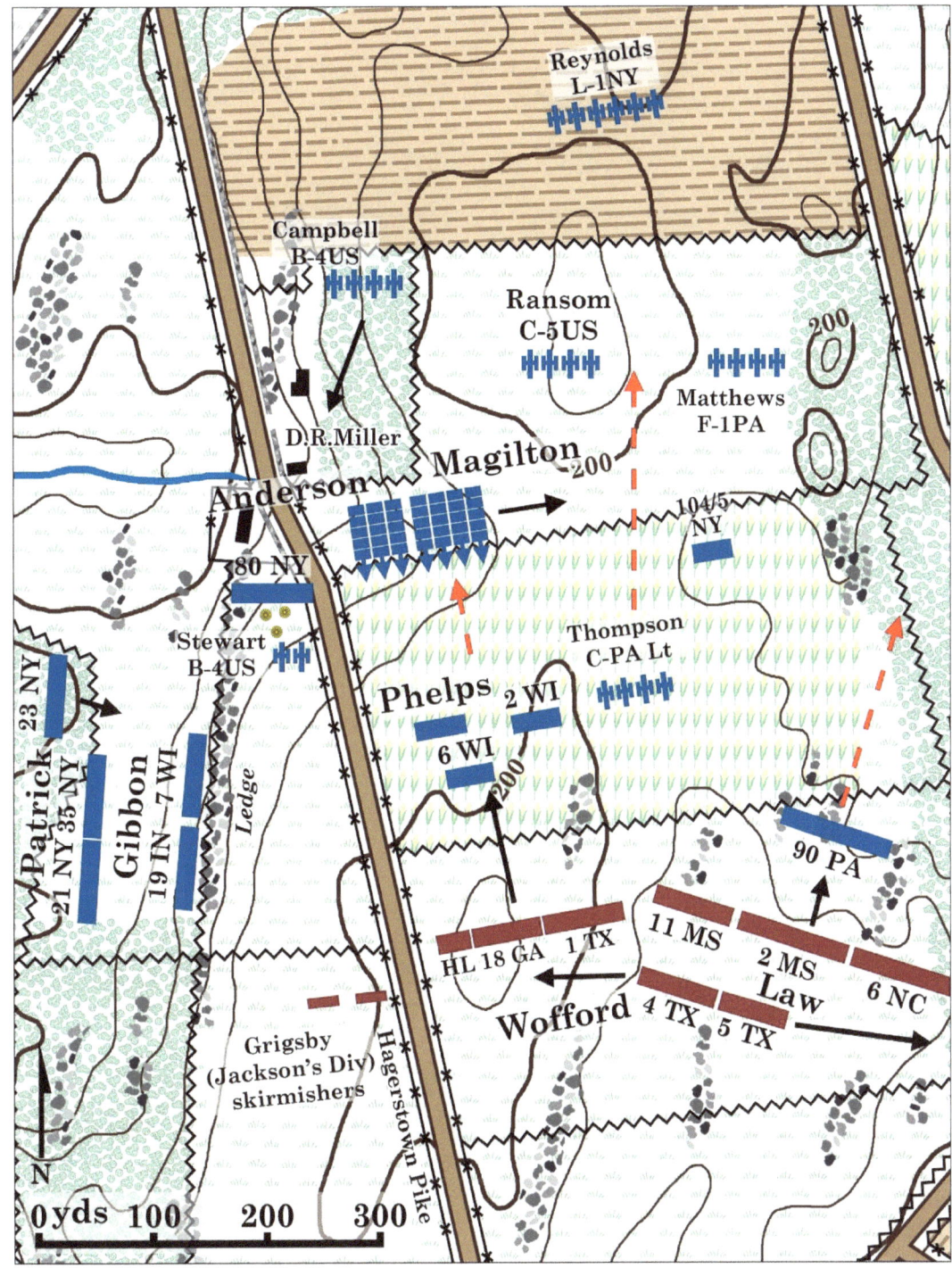

1. 7:15 a.m.
2. Hood's Texas Brigade (Wofford) pursues Gibbon and Phelps north through the Cornfield. The right of the brigade, squeezed out by Law, splits. The 4th Texas moves back to the west, while the 5th drifts east behind Law toward the East Woods.[19]
3. Magilton and Anderson's Pennsylvania Reserve brigades arrive at the Cornfield and begin deploying along the northern fence.[20]
4. The bulk of Campbell's Battery B, 4th United States moves forward to join Stewart's section in front of the 80th New York.[21]

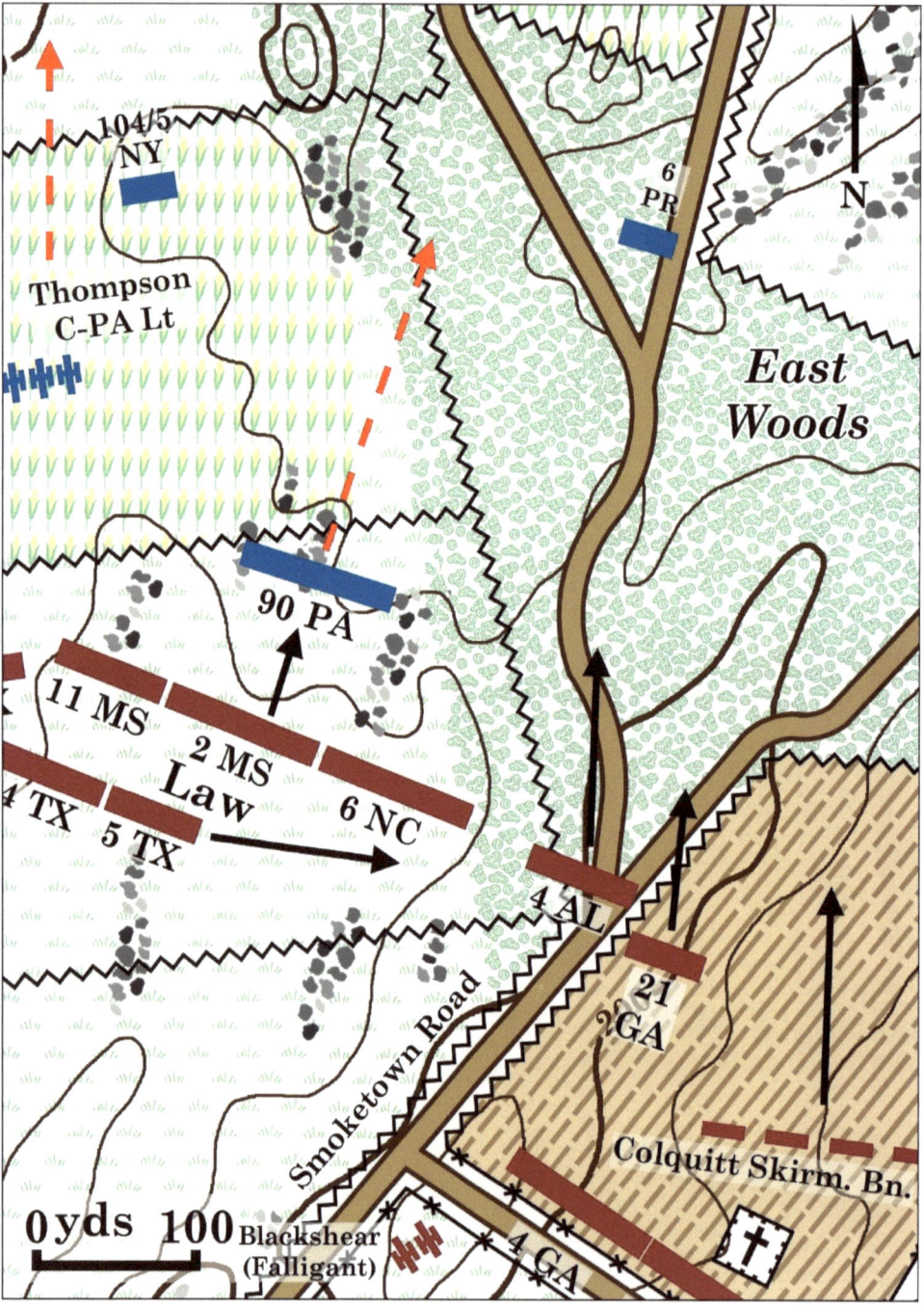

1. 7:15 a.m.
2. Law's Brigade attacks and forces the 90th Pennsylvania to retreat.[22]
3. Thompson's battery is driven from the Cornfield.[23]
4. The 5th Texas from Wofford's brigade moves to the right into the East Woods.[24]
5. The 21st Georgia, which remained behind from Trimble's Brigade, advances with Law into the East Woods. Colquitt's skirmish battalion supports their right.[25]

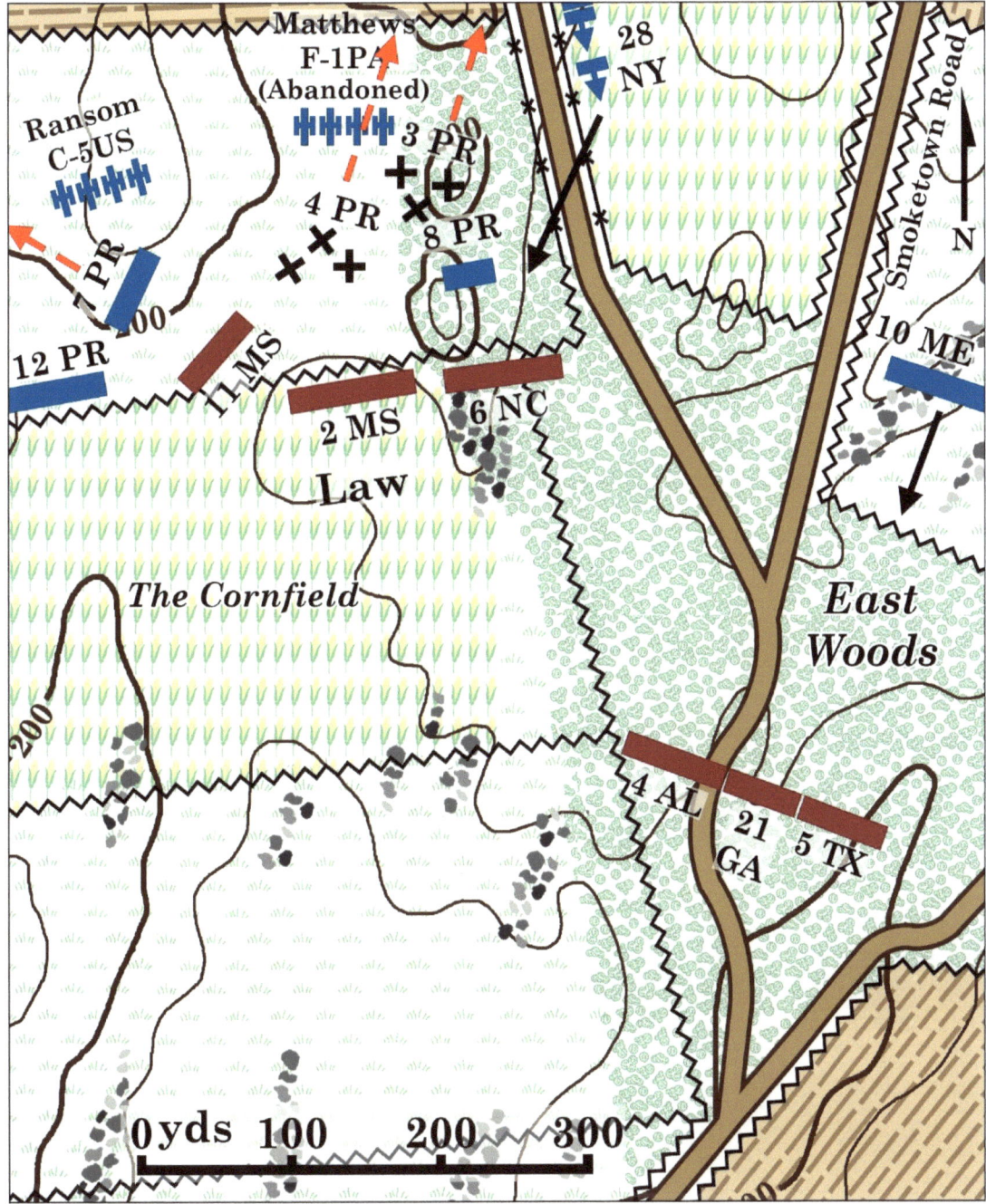

1. 7:20 a.m.
2. Laws smashes into Magilton's brigade at the northern Cornfield fence and send the 3rd and 4th Pennsylvania Reserves reeling back. The 11th Mississippi crosses the fence, forcing the 7th Reserves to retreat to the Hagerstown Turnpike and Battery F, 1st Pennsylvania to abandon their guns. The 8th Reserves fall back slightly into the northern section of the East Woods.[26]
3. The 4th Alabama, 21st Georgia, and 5th Texas form a line in the East Woods.
4. The vanguard of the Twelfth Corps, the 10th Maine of Crawford's brigade, arrives and advances to the fence at the northern edge of the East Woods. Tyndale's brigade is not far behind, coming up behind Magilton and the 8th Reserves.[27]

The Twelfth Corps Attacks

With the arrival of Brigadier General Joseph K. Mansfield's Twelfth Corps the weight of the Union juggernaut proved an insurmountable advantage. Hood's Division was attacked on three sides by the rest of Doubleday's division, Meade's Pennsylvania Reserves, and the new arrivals. The weight in numbers forced Hood back into the West Woods.

However, the Union weren't the only ones receiving reinforcements. D. H. Hill continued to send additional brigades into the fight. Brigadier General Roswell S. Ripley's brigade, which had been supporting Lawton, advanced and stopped the Union pursuit of Hood's retreating men. Brigadier General Alfred H. Colquitt then led his brigade past Ripley and into the cornfield, once again driving the Union back to the northern fence. Once again, Union numbers proved their undoing. Brigadier General Alpheus S. William's Twelfth Corps division poured into the East Woods and around Colquitt's right flank. General Mansfield was mortally wounded leading and coordinating his men. The pressure from front and right proved too much, and Colquitt, Ripley, and a supporting brigade were forced to retreat all the way back to the West Woods. General Hooker was wounded while leading the advance and removed from the field.

To the west of the Hagerstown Turnpike Colonel William B. Goodrich led his brigade past the Miller farm into the northern extension of the West Woods. He asked Brigadier Marsena R. Patrick to advance with him, but Patrick demurred and sought reinforcements. While he was away, Goodrich led the two brigades forward, his brigade in the lead. Soon after advancing they met Brigadier General Jubal A. Early's Brigade and were stopped cold. Colonel Goodrich was killed and the two sides continued their fight.

Even more reinforcements were on the way for both sides.

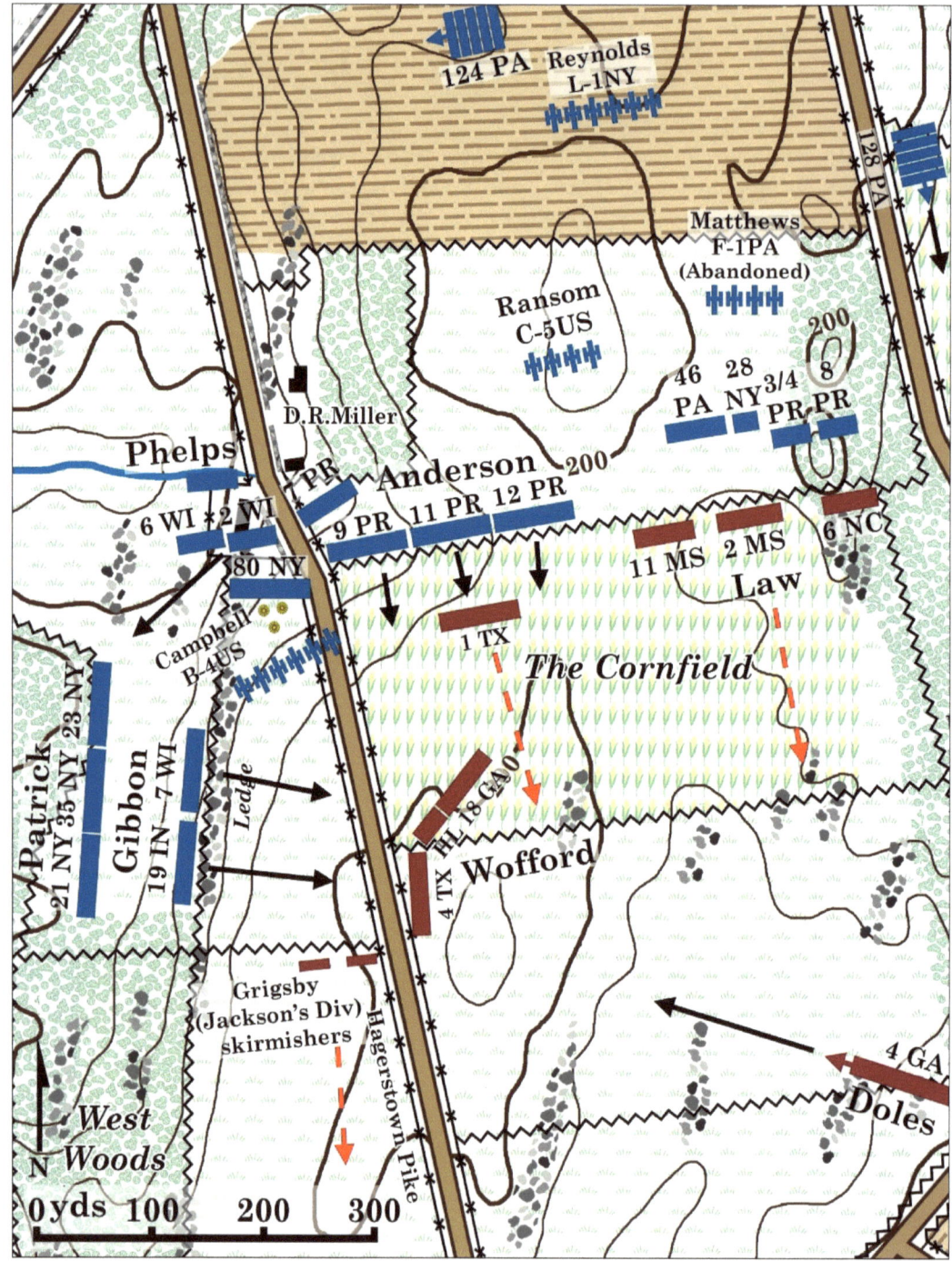

1. 7:30 a.m.
2. The 1st Texas charges to the northern edge of the Cornfield, but are devastated by Anderson's line of Pennsylvania Reserves and Battery B, 4th United States. They retreat back south. Anderson pursues.[1]
3. Fire from Battery B stops the 4th Texas, Hilliard Legion, and 18th Georgia. Wofford also needs to contend with Gibbon's advancing right wing, supported by Patrick's brigade. Gibbon's advance forces them to retreat, along with Grigsby's skirmishers.[2]
4. The 3rd and 4th Reserves rally and move back to the front. Crawford's 46th Pennsylvania and 28th New York deploy and seal off the gap. Law's Brigade retreats after taking heavy casualties.[3]
5. Ripley's Brigade (under Doles) leaves Mumma's lane and marches into the field south of the Cornfield.

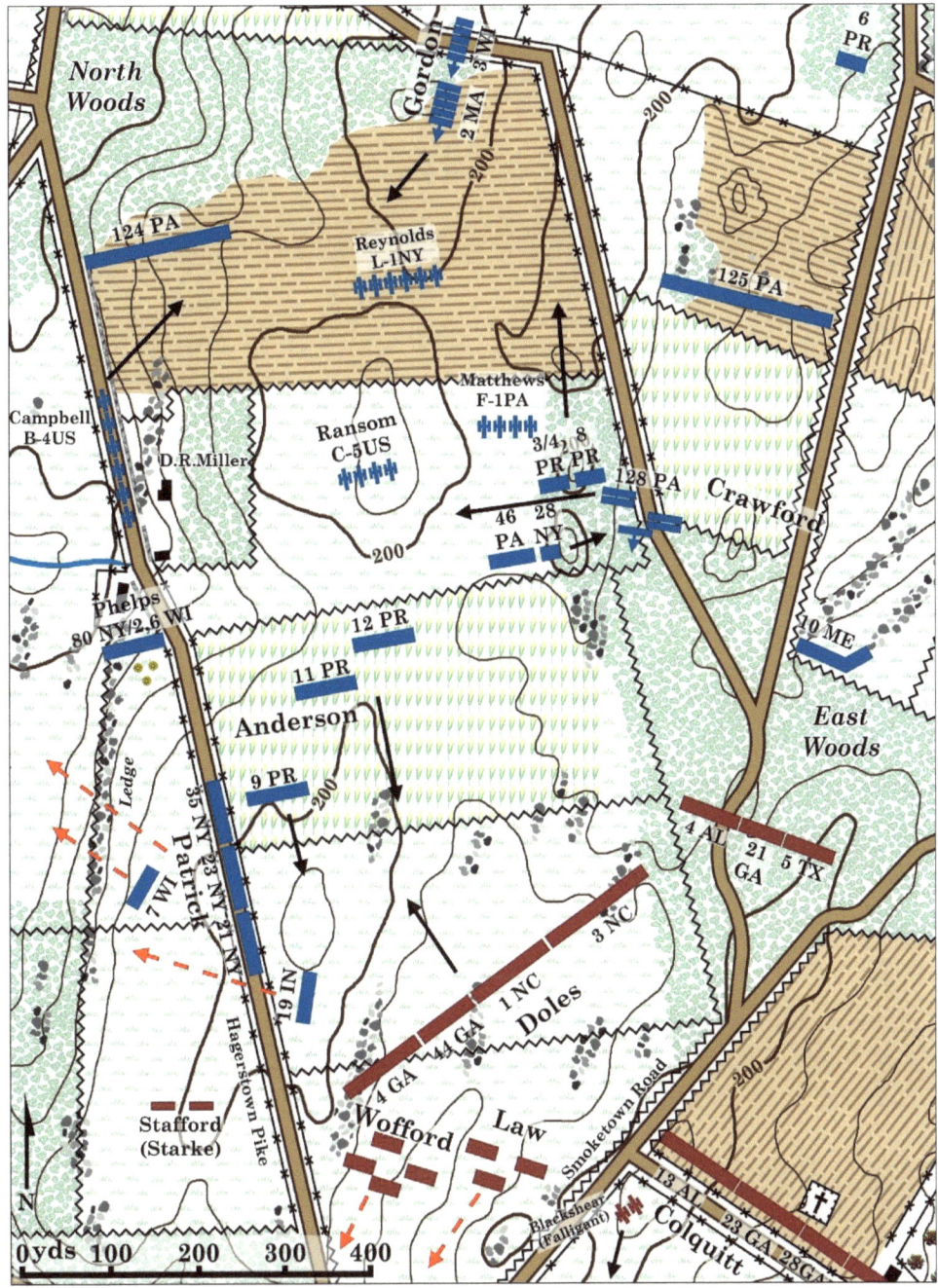

1. 7:40 a.m.
2. Doles forms into line and advances north. Gibbon's right wing and Patrick retreat past the limestone ledge. About 100 men from Starke's brigade, commanded by Colonel Leroy A. Stafford, advance alongside on the left of the Hagerstown Turnpike and fire into their flank.[4]
3. Anderson pursues the retreating Texas Brigade and runs into Doles at the southern edge of the Cornfield.
4. Magilton's brigade retires from the field.
5. The inexperienced 128th Pennsylvania loses formation attempting to form into line under fire. Its colonel is killed and lieutenant colonel wounded. Colonel Joseph F. Knipe of the 46th Pennsylvania attempts to sort out the confusion.[5]
6. Colquitt's Brigade takes up a position in the Mumma lane after shifting north from the Sunken Road.

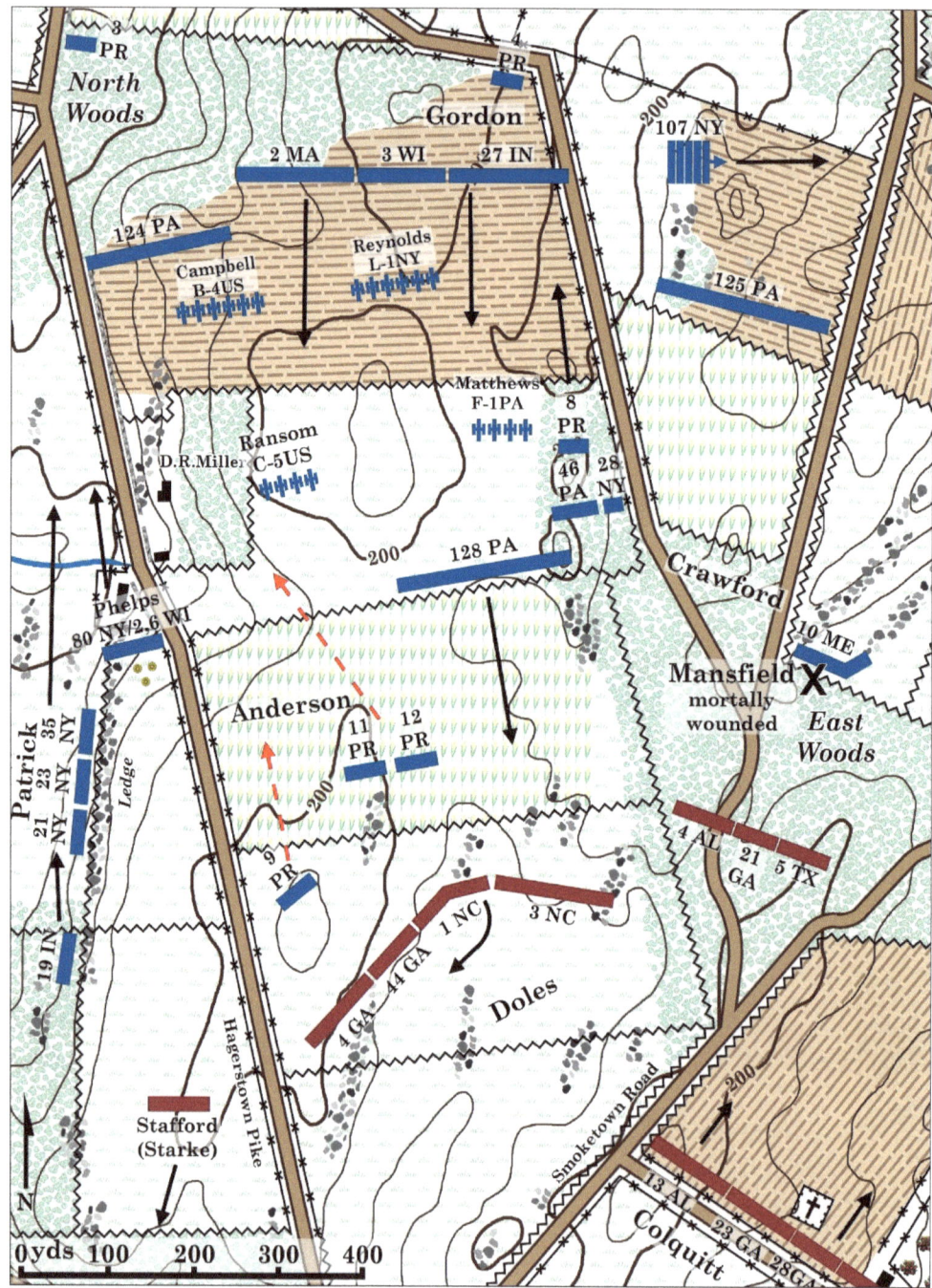

1. 7:45 a.m.
2. Doles fights Anderson at the southern edge of the Cornfield. Outnumbered, Anderson retreats.
3. The 1st North Carolina splits, a portion moving to the left of the line.[6]
4. The 128th Pennsylvania charges through the Cornfield toward Doles, taking heavy casualties.[7]
5. Thinking that the 10th Maine is firing into friendly units, Mansfield rides along the regiment ordering them to cease fire. He is mortally wounded as the men are pointing out his error.[8]
6. Patrick and Gibbon retire from their advanced position near the limestone ledge.
7. Gordon's brigade forms into line of battle north of the Cornfield and advances to the high ground overlooking the field.
8. Colquitt advances toward the East Woods.

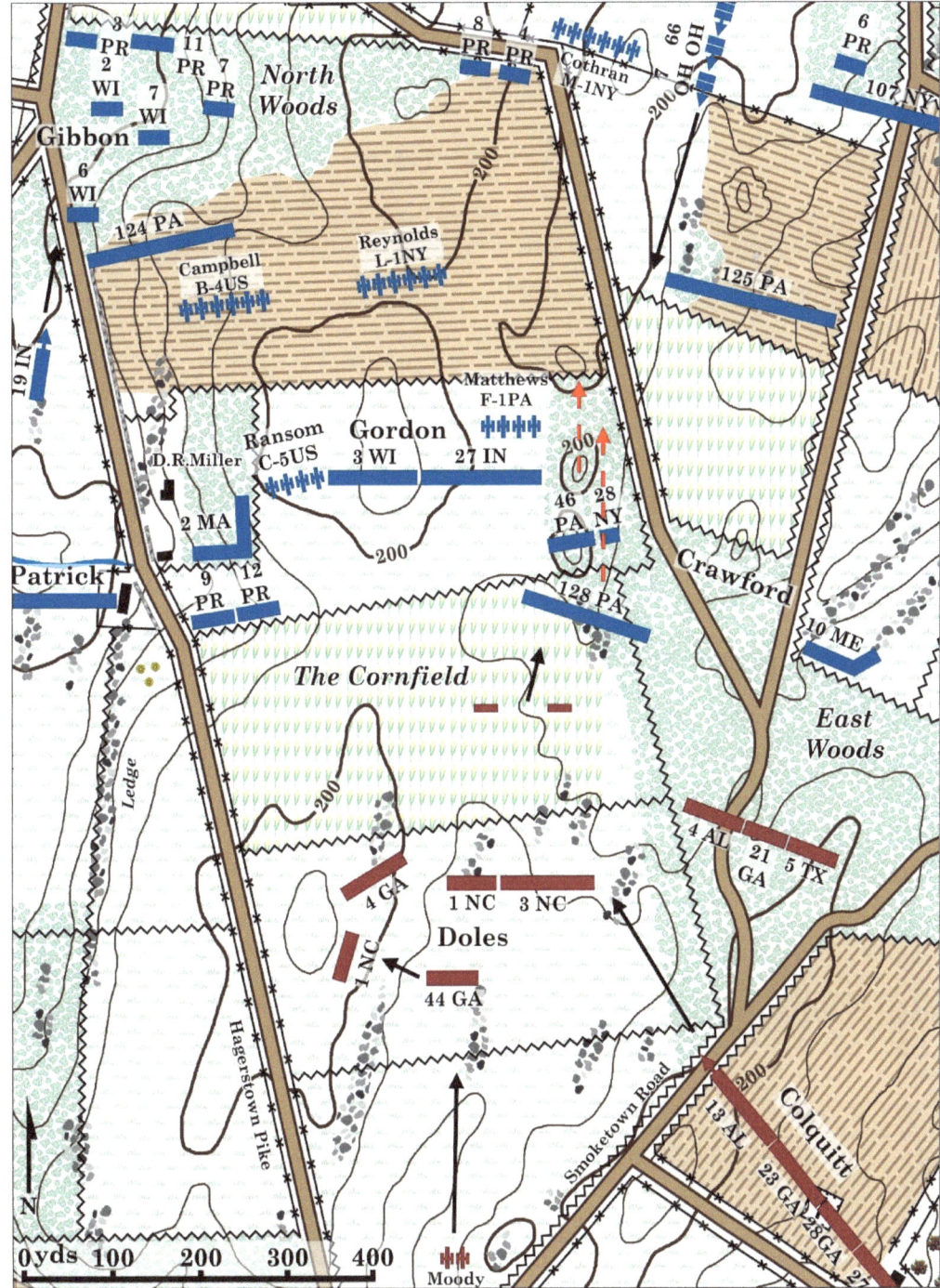

1. 8:00 a.m.
2. Gordon forms on the high ground north of the Cornfield and engages Doles in a long range firefight. He is supported by Battery C, 5th United States and Battery F, 1st Pennsylvania. Both sides suffer severely.[9]
3. Crawford's three regiments at the north end of the East Woods retreat.
4. Gibbon, Magilton, and Anderson reform in the North Woods. The 9th and 12th Reserves remain at the fence south of the Miller house.
5. The advanced units of Greene's division arrive on the field.
6. Shortly after advancing from Mumma lane, Colquitt to moves to the left to support Doles.

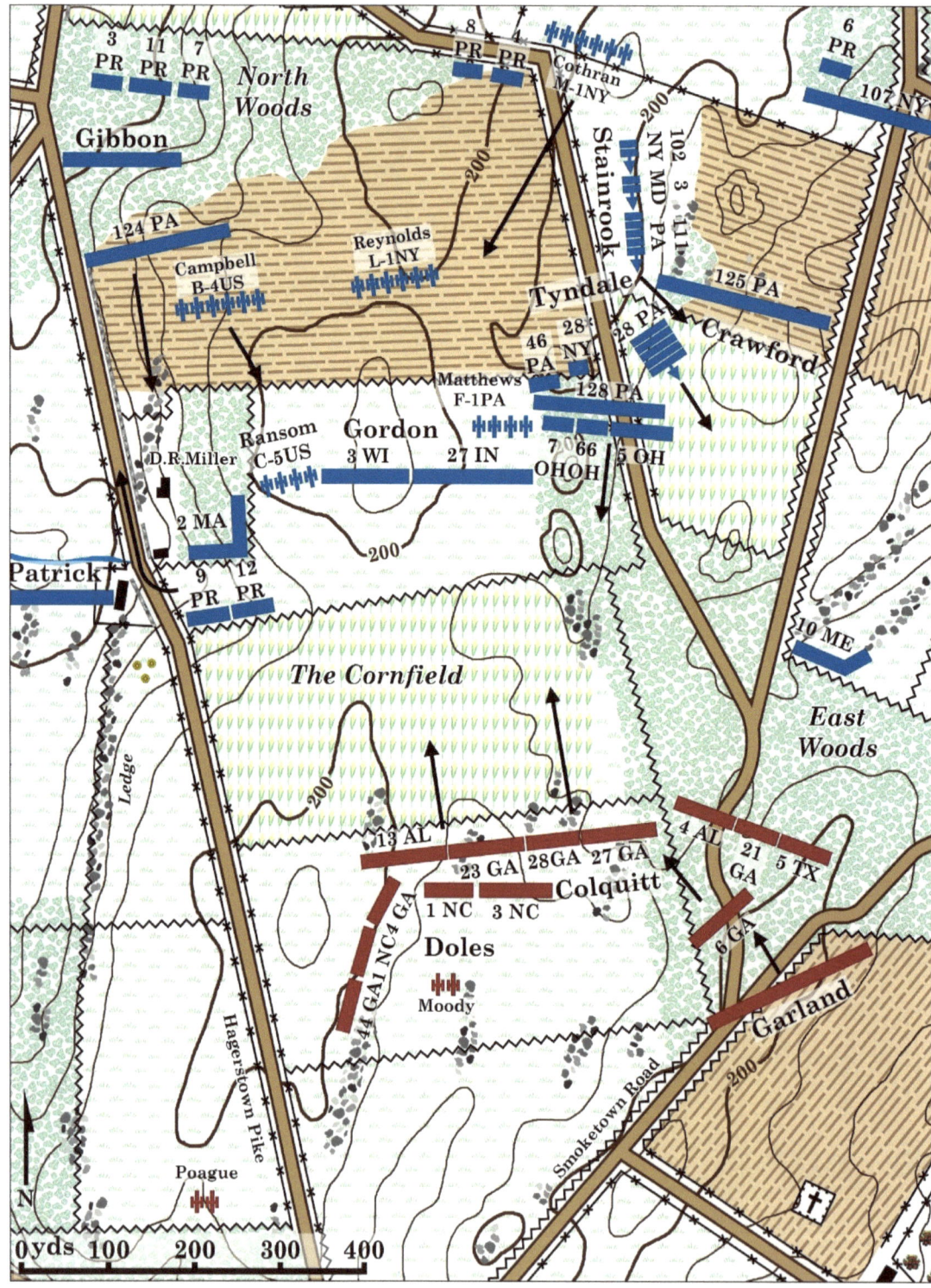

1. 8:15 a.m.
2. Colquitt marches in front of Doles and moves north into the Cornfield.
3. Garland's Brigade moves north into the East Woods.
4. Tyndale and Stainrook's brigades from Greene's division deploy and move toward the northeast corner of the Cornfield.
5. The 124th Pennsylvania advances south down the Hagerstown Turnpike.[10]

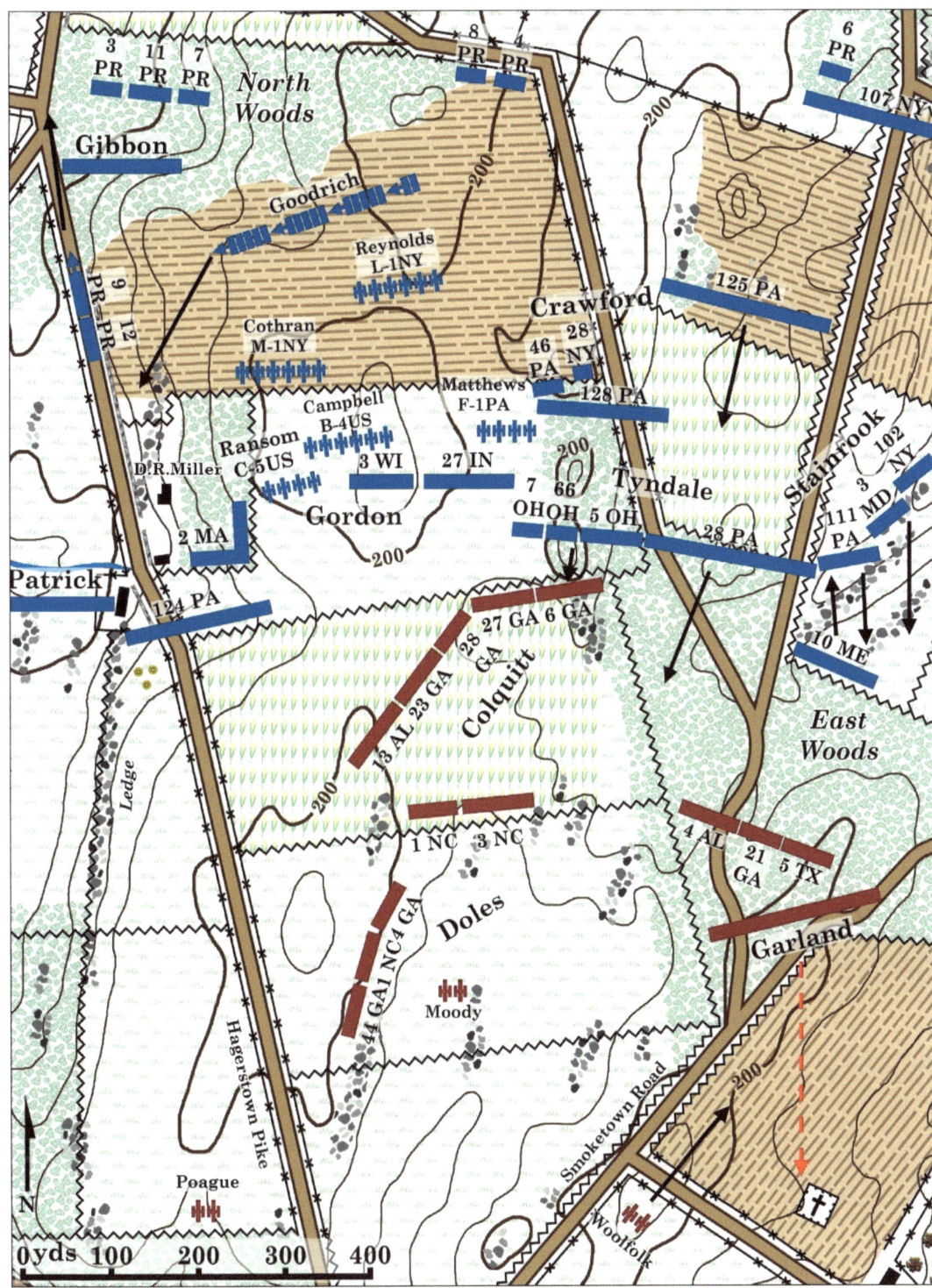

1. 8:30 a.m.
2. Tyndale advances and meets the 6th and 27th Georgia at point blank range across the Cornfield fence. The 28th Pennsylvania begins moving south through the East Woods to flank Colquitt.[11]
3. Stainrook arrives and relieves the 10th Maine.[12]
4. Goodrich moves across the fields towards the Hagerstown Turnpike.
5. Garland's Brigade, roughly handled at South Mountain just days before, melts away under fire from the Federals in the East Woods.[13]

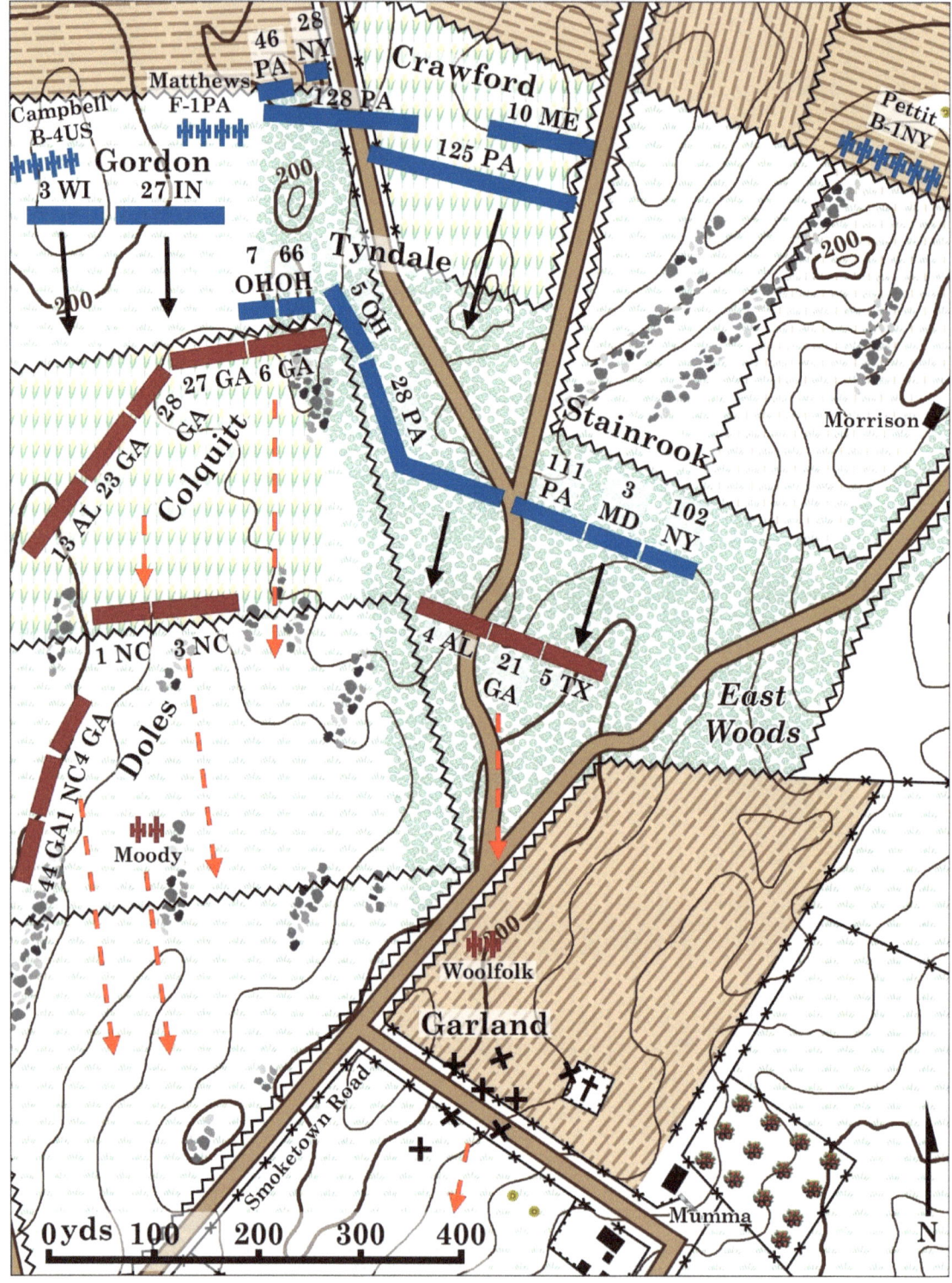

1. 8:40 a.m.
2. The 5th Ohio and 28th Pennsylvania flank Colquitt's line. The Ohioans unleash a devastating volley into the 6th Georgia at only a few yards distance, decimating the right-hand companies.[14]
3. Gordon's brigade, elements of Crawford's brigade, and Greene's division advance south. The onslaught is too much for the Confederates, and Colquitt, Doles, and the three regiments in the East Woods retreat.

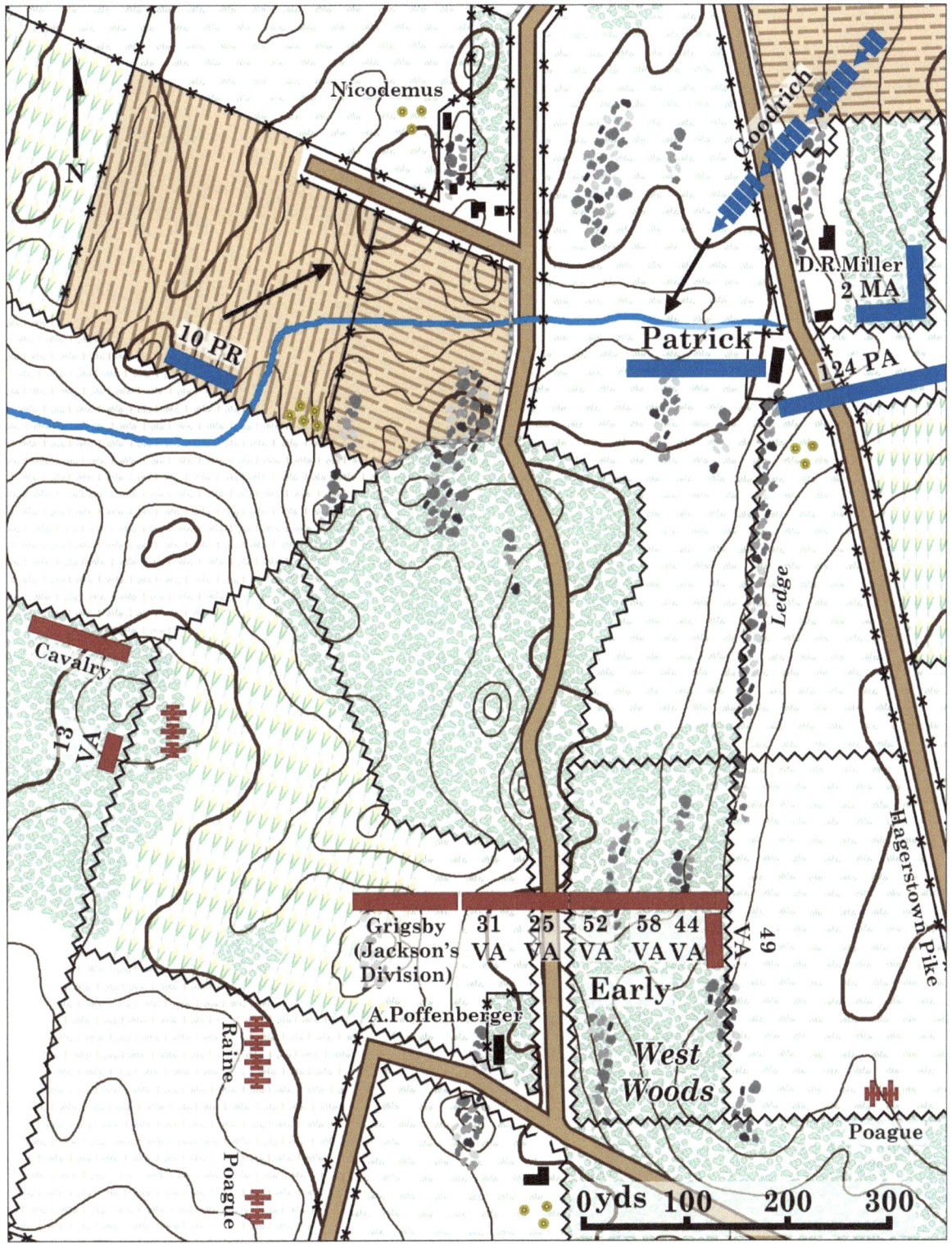

1. 8:45 a.m.
2. Goodrich's brigade moves across the Hagerstown Turnpike. Goodrich forms in front of Patrick with the intention of entering the West Woods to confront Early's Brigade.
3. Patrick goes across the turnpike to find reinforcements, but returns empty-handed. He finds that Goodrich and his brigade have advanced in his absence.[15]

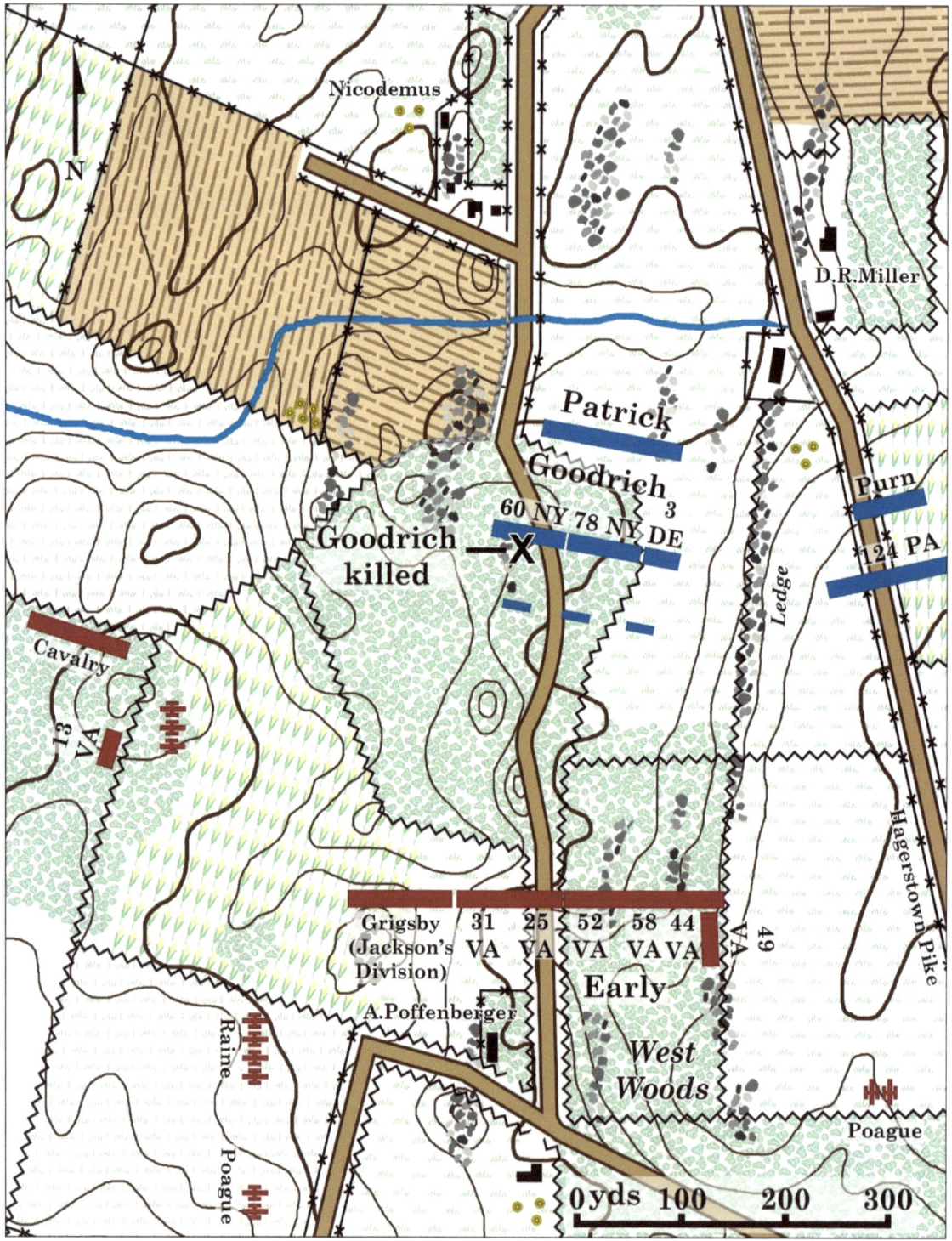

1. 8:50 a.m.
2. Colonel Goodrich advanced his brigade into the northern section of the West Woods. Patrick's brigade follows in support.
3. The brigade takes fire from Early's Brigade and the remnants of Jackson's Division.
4. Goodrich is killed near the 60th New York, his old command, as the regimental adjutant is nearby and reaches him quickly.[16]
5. The 124th Pennsylvania and Purnell Legion advance down the Hagerstown Turnpike.[17]

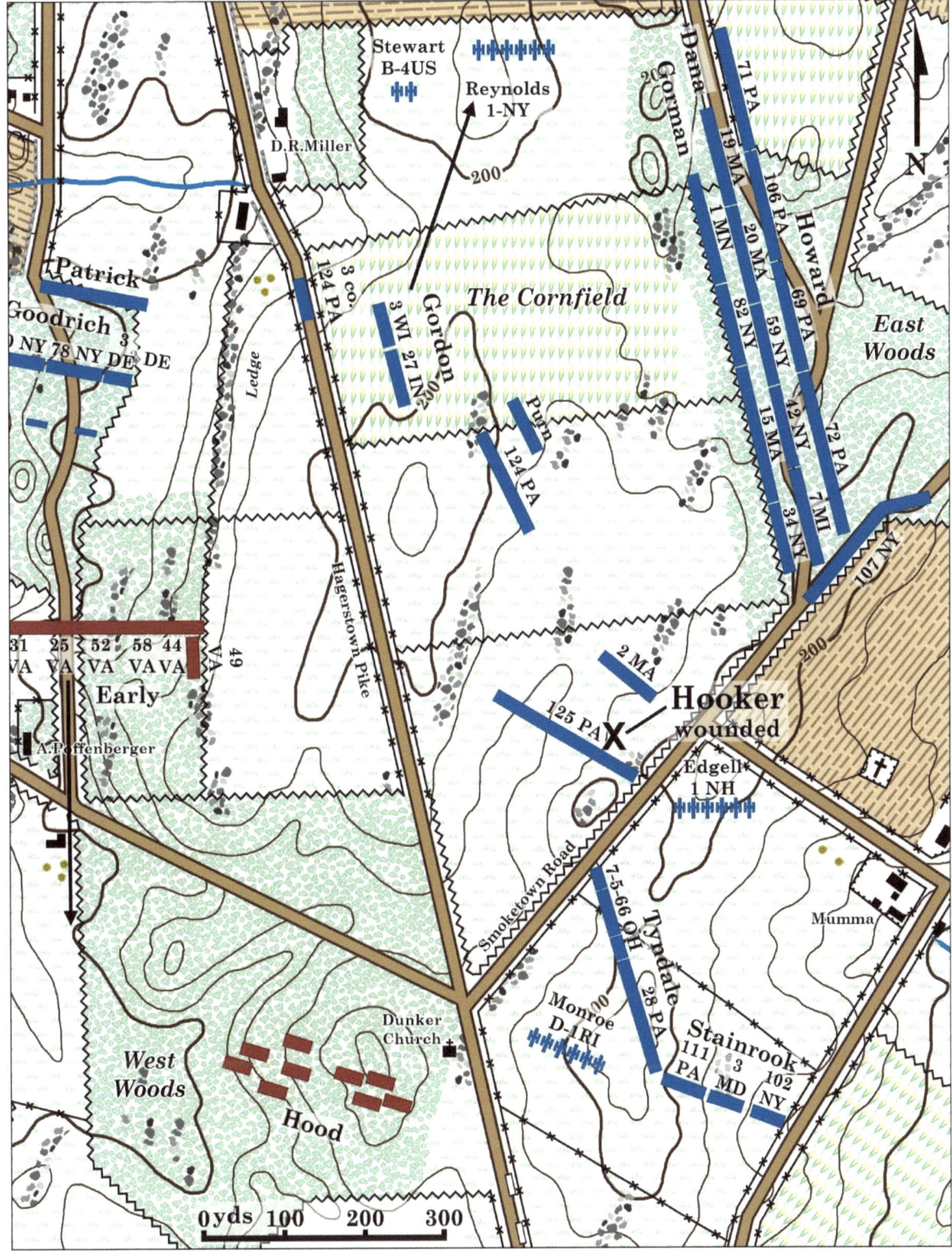

1. 8:55 a.m.
2. The Second Corps arrives in the East Woods.
3. The 124th Pennsylvania and Purnell Legion move east into the Cornfield, except three companies of the 124th that remain at the Hagerstown Turnpike.

4. Gordon's brigade falls back north of the Cornfield.
5. Tyndale and Stainrook pursue the defeated Confederates fleeing the East Woods and form on the plateau opposite the Dunker Church. Battery D, 1st Rhode Island unlimbers and duels with Confederate artillery to the south.[18]
6. Early judges the threat to his rear at Dunker Church to be greater than Goodrich and Patrick to his front, as the two brigades are not advancing. Early marches his brigade south to confront the danger there.[19]
7. General Hooker is likely wounded talking to Colonel Jacob C. Higgins of the 125th Pennsylvania, and rides off. He later reels in the saddle from loss of blood and is carried from the field.[20]

The West Woods

Brigadier General George S. Greene led his Twelfth Corps division forward and established a reverse slope defense on the plateau across the pike from the Dunker Church. William's division from the same corps was strung out along the battlefield. Only one of William's isolated regiments occupied the West Woods around the church.

Major General Edwin V. Sumner's Second Corps arrived as Greene pushed to the Dunker Church. After conducting a reconnaissance, Sumner devised a plan to push back the Confederates and move south. Major General John Sedgwick would march west into the West Woods, push back Early's Brigade confronting Goodrich, and then wheel south with Green's division anchoring his left. Brigadier General William H. French would move his division south and form up with Greene's division on his right. Then Major General Israel B. Richardson's division would support both as necessary.

Sedgwick moved out, crossed the Miller farm and Cornfield, and entered the northern reaches of the West Woods. Unfortunately for him, a large volume of Confederate reinforcements arrived at the same time. First to arrive was Major General Lafayette McLaw's division, followed immediately by Brigadier General John G. Walker's division. Sedgwick was still oriented north to south when five Confederate brigades slammed into his left flank from the shelter of the woods. Stacked up one behind the other, his three brigades had no room to maneuver. Sedgwick's men soon took flight and had to rally in the East and North Woods.

The fight continued for the West Woods as the battle raged around the Sunken Road. After Sedgwick was pushed back, Greene advanced his division into the relatively empty woods and established a defensive perimeter around the church. As the Sunken Road position was collapsing Walker's Division launched a counterattack that drove Greene from the woods. Two regiments continued onward and cleared the plateau across the turnpike, then continued east to take French's rallying command in flank. The federals quickly gathered and presented a stout defense.

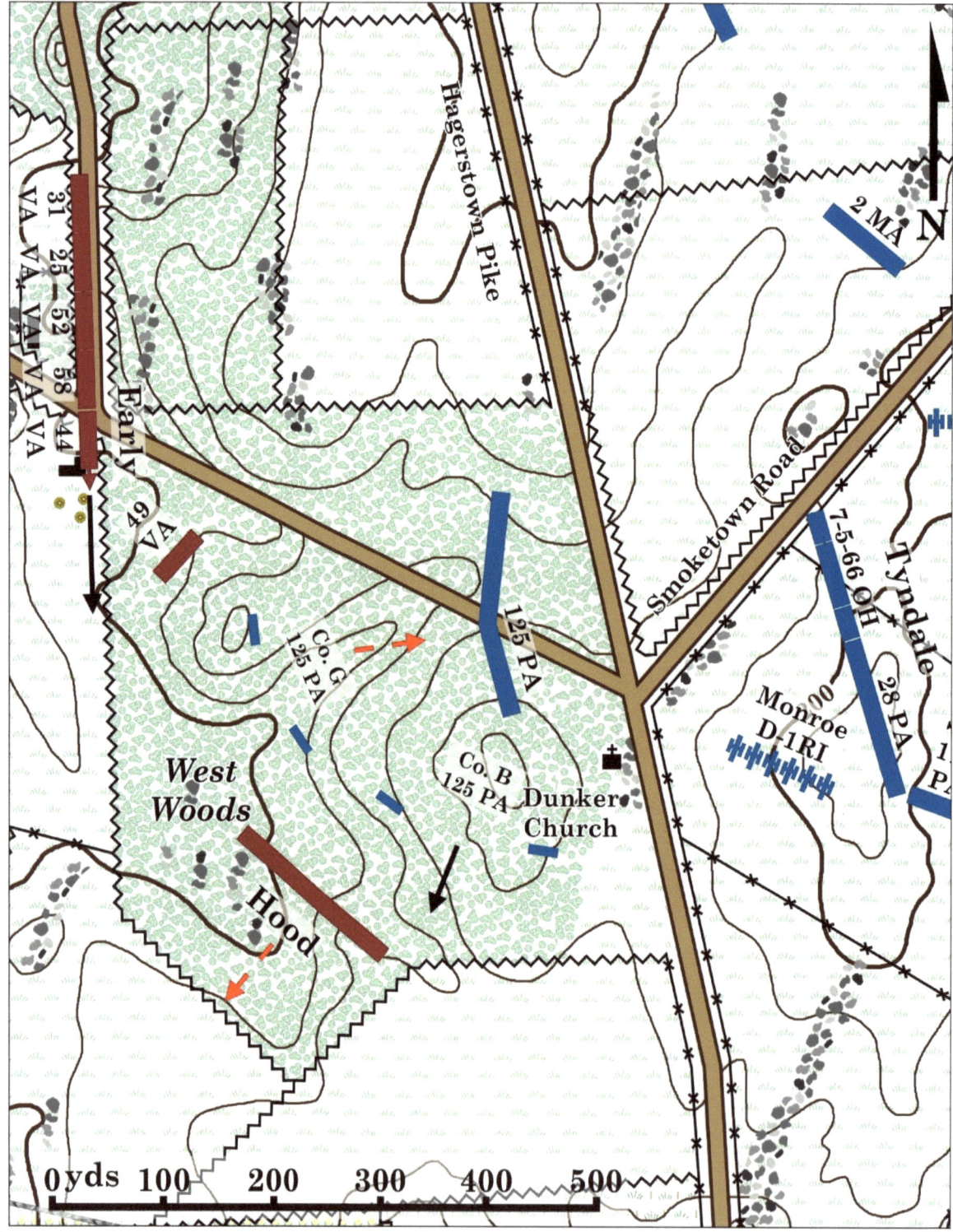

1. 9:00 a.m.
2. The 125th Pennsylvania is ordered into the West Woods. They enter north of the Smoketown Road and form a line on high ground. Two companies deploy as skirmishers. Company G runs into the 49th Virginia of Early's Brigade and falls back. Company B pursues the remnants of Hood's Division to the south edge of the West Woods.[1]

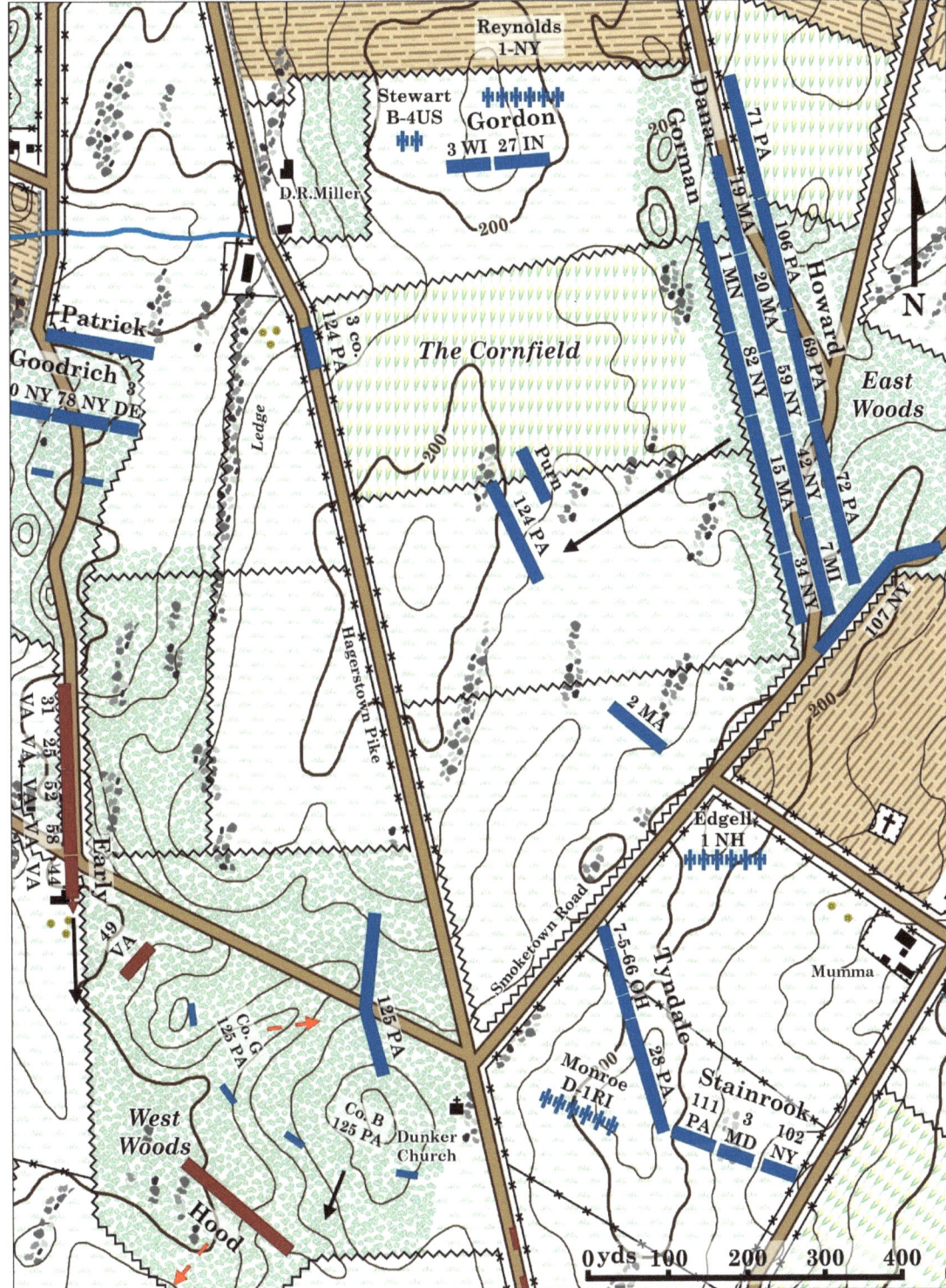

1. 9:00 a.m.
2. Sedgwick's division begins its advance across the Cornfield and towards the West Woods. Second Corps commander Major General Edwin V. Sumner's intent is to have Sedgwick's Second division gain the West Woods and turn south, with Green's division on his left. Then French's Third division will form on Greene's left, presenting a solid line facing southward towards the Confederates.[2]
3. The brigades of the division are lined up in close column one brigade behind the other.

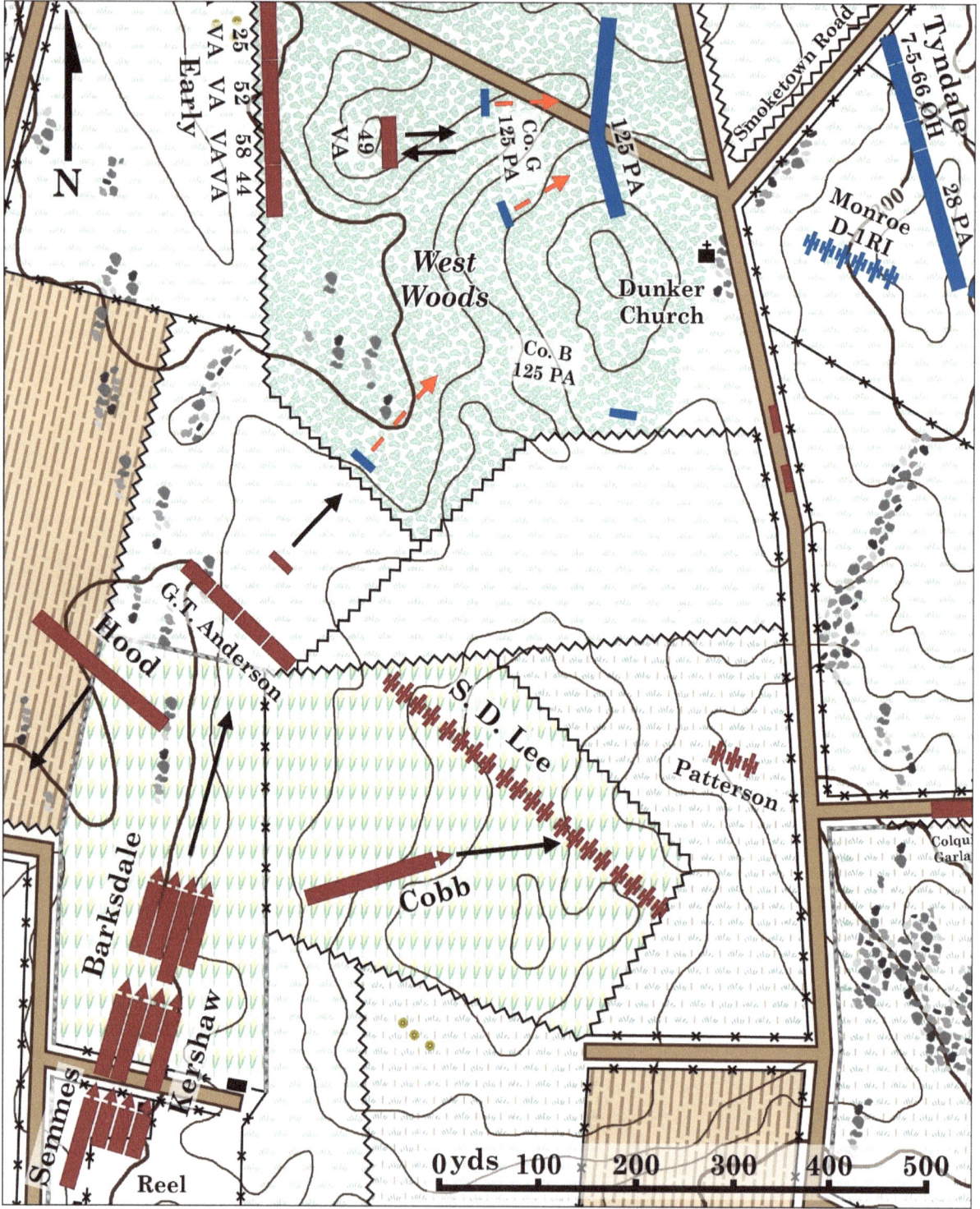

1. 9:05 a.m.
2. The 49th Virginia advances but falls back.³
3. McLaw's Division approaches the West Woods from the south. Cobb's Brigade is ordered east to the Sunken Road.⁴
4. Skirmishers from George. T. Anderson's brigade advance and drive off the 125th Pennsylvania skirmishers from the fence on the southern border of the West Woods.⁵

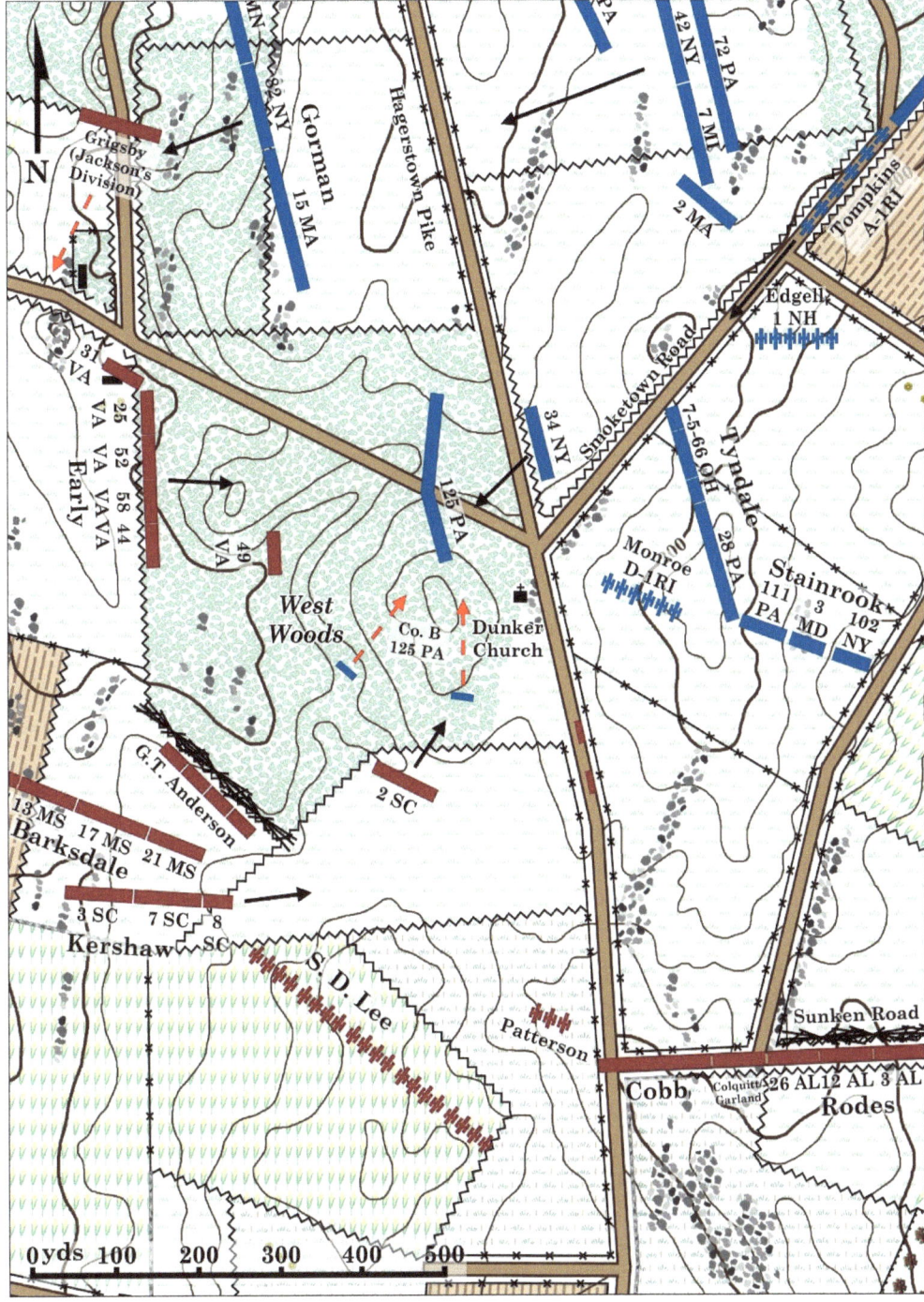

1. 9:10 a.m.
2. Gorman's brigade advances through the northern portion of the West Woods and forces the remaining men of Jackson's Division, who are still fighting Goodrich and Patrick, to retreat.
3. The 34th New York from Gorman's brigade advances into the West Woods to support the 125th Pennsylvania.[6]
4. G. T. Anderson's Brigade tears down the fence bordering the West Woods and begin to build a barricade.[7]
5. Early's Brigade and the 2nd South Carolina of Kershaw's Brigade advance toward the 125th Pennsylvania.

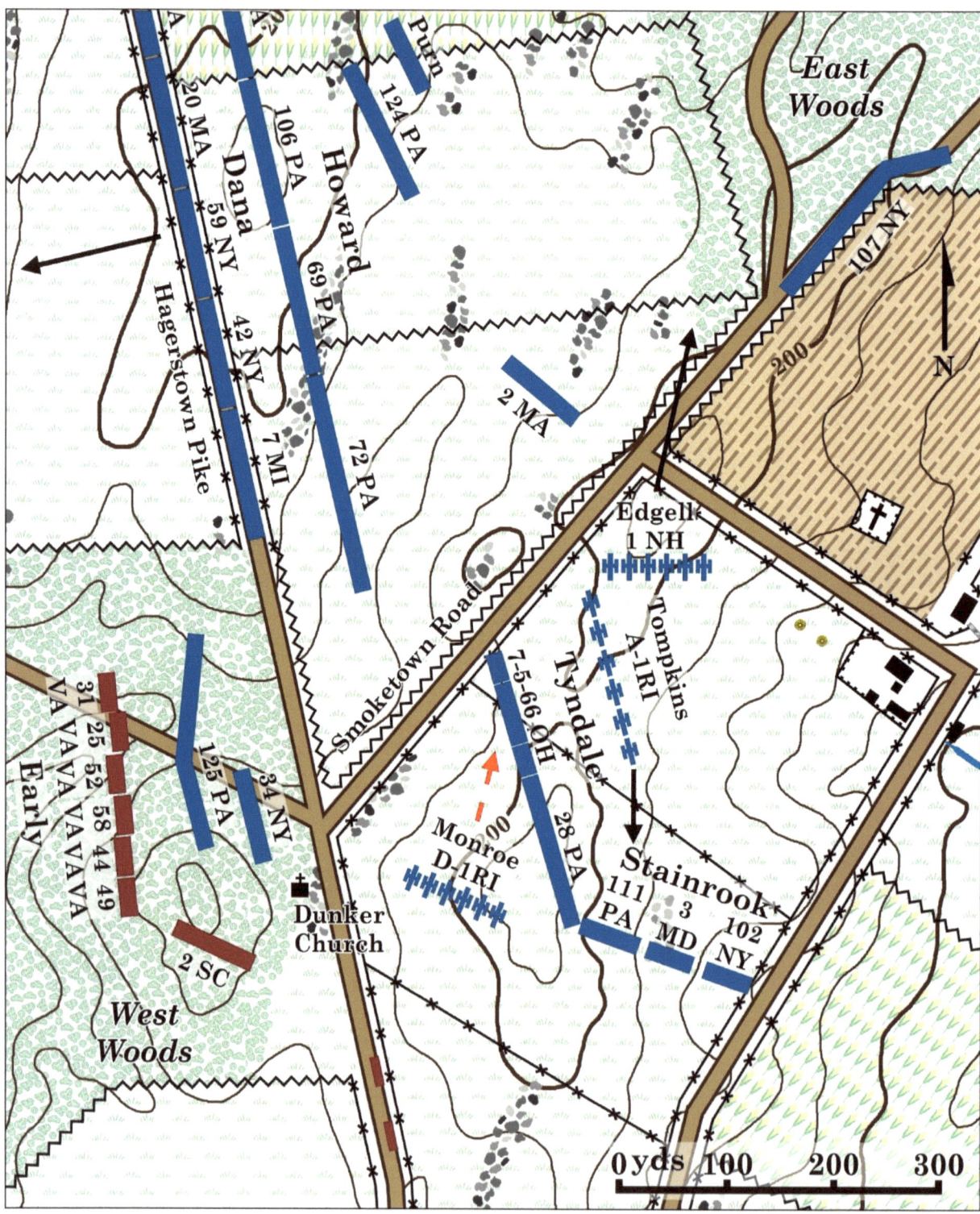

1. 9:15 a.m.
2. Early and the 2nd South Carolina engage the 125th Pennsylvania and the 34th New York.[8]
3. Battery D, 1st Rhode Island takes heavy casualties from Confederate skirmishers sheltered in the Hagerstown Turnpike to the southwest. They are forced to limber up and retire.[9]
4. Battery A, 1st Rhode Island unlimbers to the southeast of Battery D's position, taking advantage of the hilltop to provide better cover.[10]

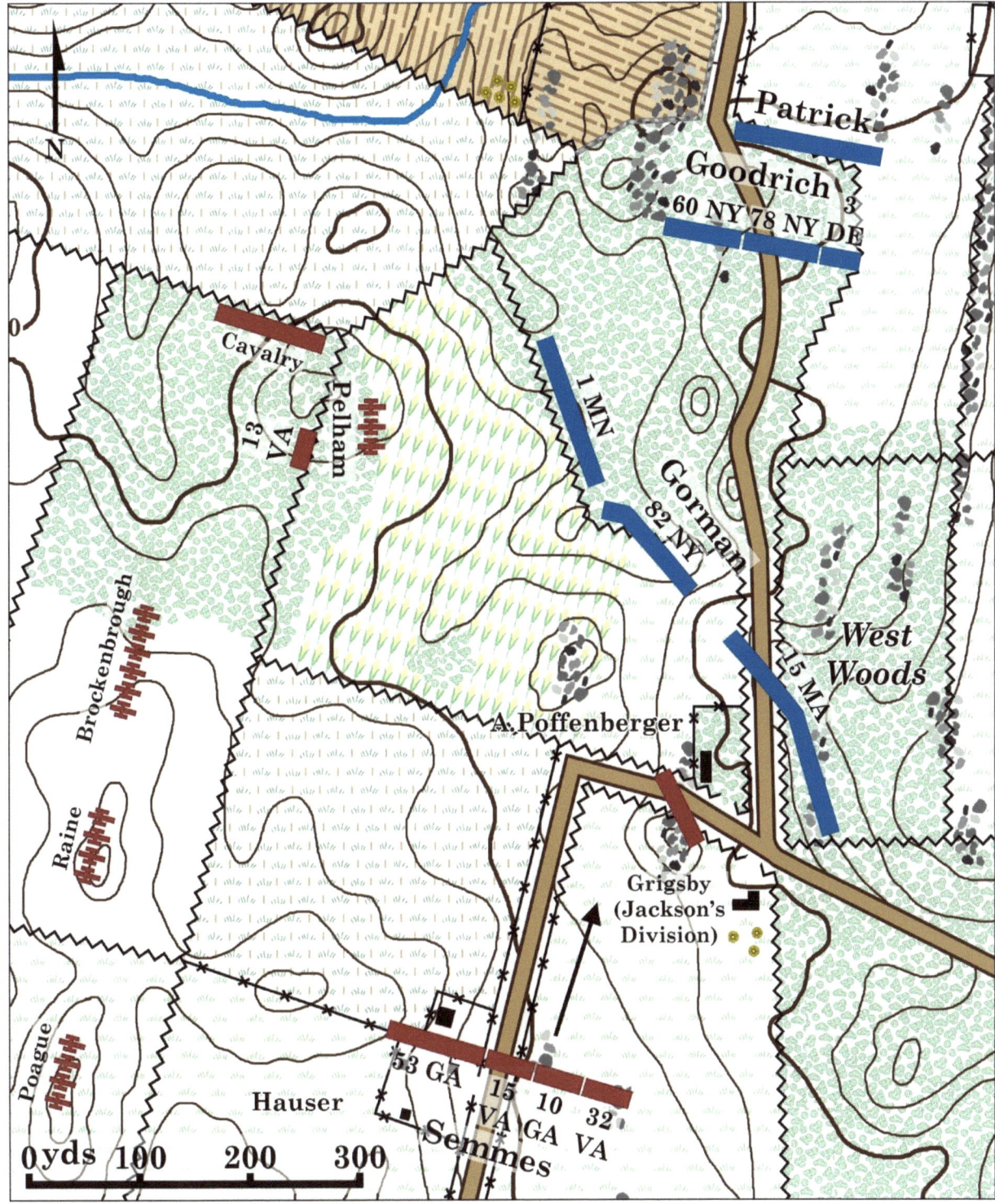

1. 9:20 a.m.
2. Gorman's brigade passes in front of Goodrich and stops at the western fence of the West Woods. The brigade takes fire from several Confederate batteries to the west.
3. The 15th Massachusetts engages Jackson's Division.[11]
4. Semmes Brigade arrives and moves north to confront Gorman.

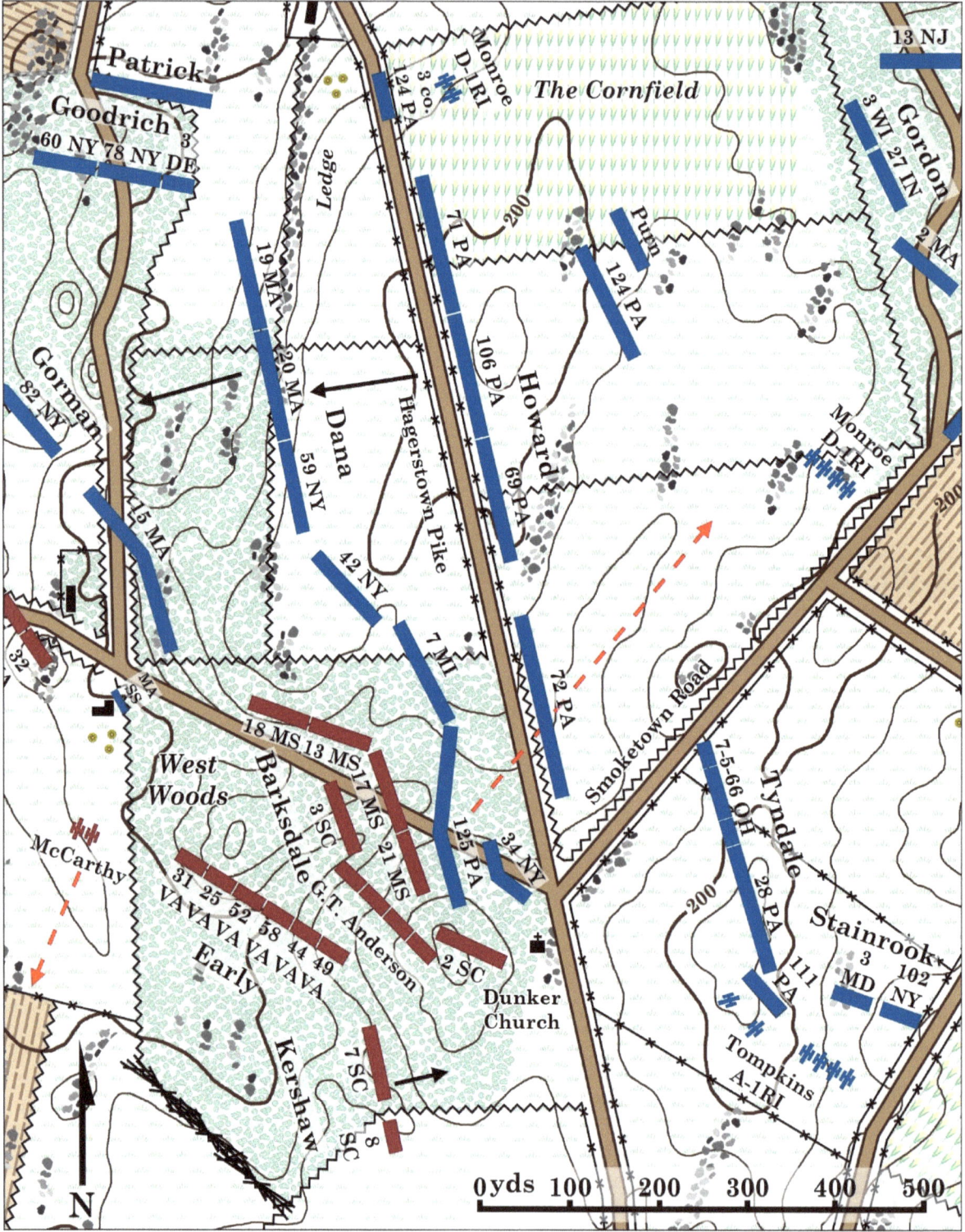

1. 9:25 a.m.
2. Dana and Howard continue to move west, but their left regiments begin to encounter resistance at the West Woods and have to turn to face this new threat.
3. Barksdale and G. T. Anderson advance through Early and attack the 125th Pennsylvania. The regiment finally breaks and falls back out of the woods.[12]

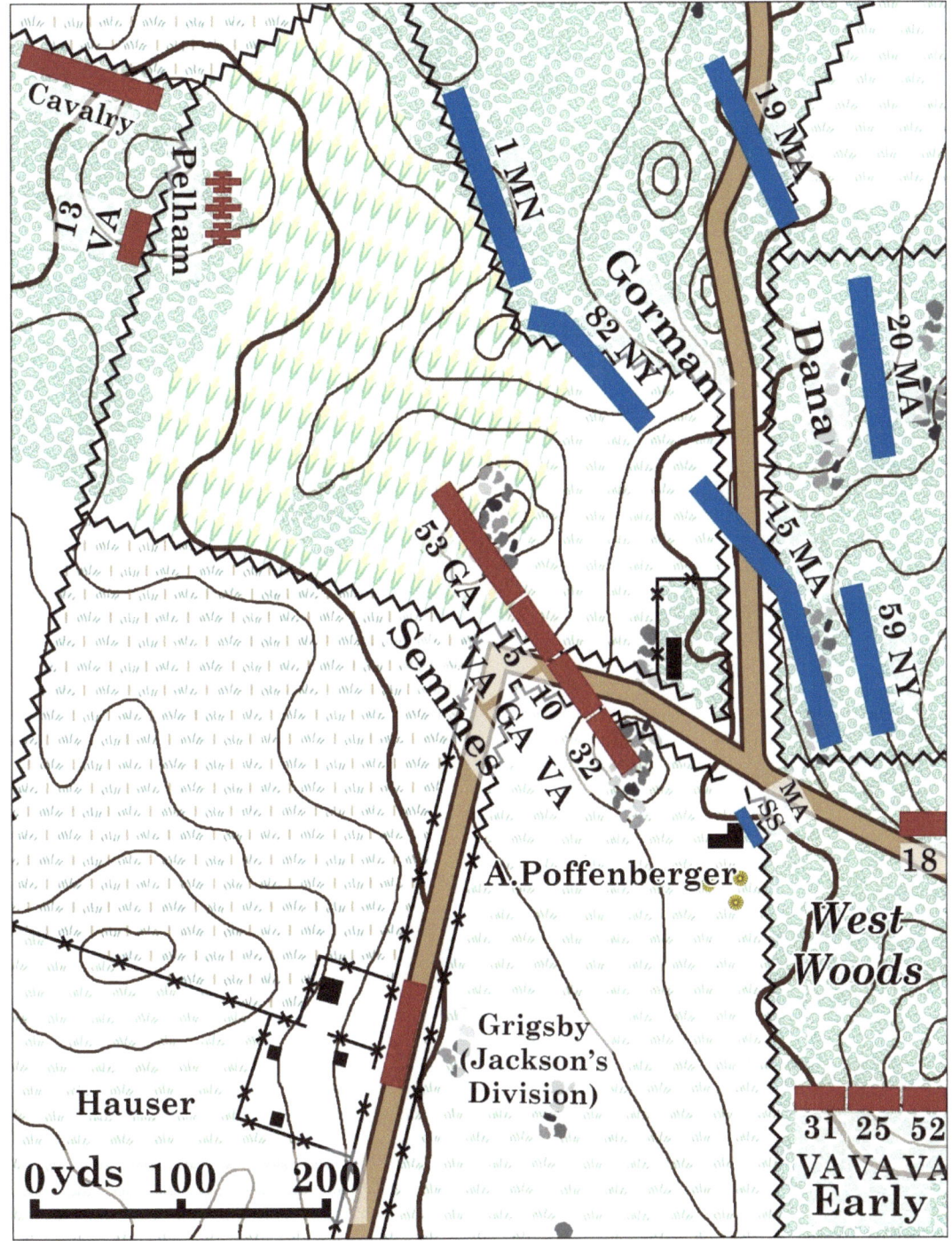

1. 9:30 a.m.
2. Semmes' Brigade settles down into a firefight with Gorman.[13]
3. Jackson's Division takes shelter in the Hauser farm lane, and trades shots with the Massachusetts sharpshooters at the Poffenberger barn.

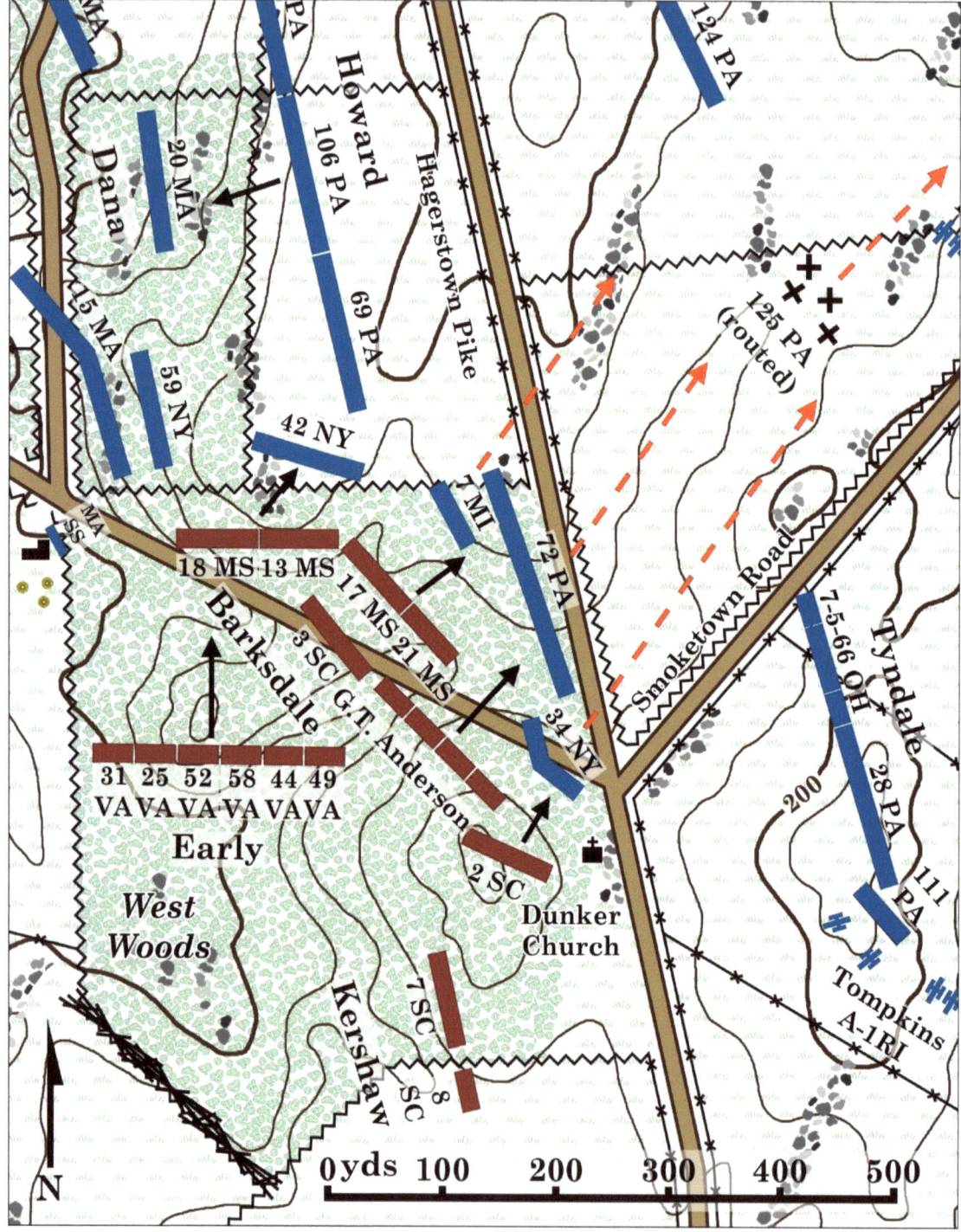

1. 9:30 a.m.
2. McLaws' momentum carries the division north and northeast. The pressure from Kershaw, G. T. Anderson, and Barksdale's right wing is too much for the Union regiments remaining, and the 7th Michigan, 72nd Pennsylvania, and 34th New York retire from the woods.[14]
3. The left of Barksdale's Brigade engages the 42nd New York as they attempt to swing south to meet the unexpected threat.[15]
4. Howard, unaware of the threat, continues to move west. The brigades of Sedgwick's division are now crowded together with little room to maneuver.

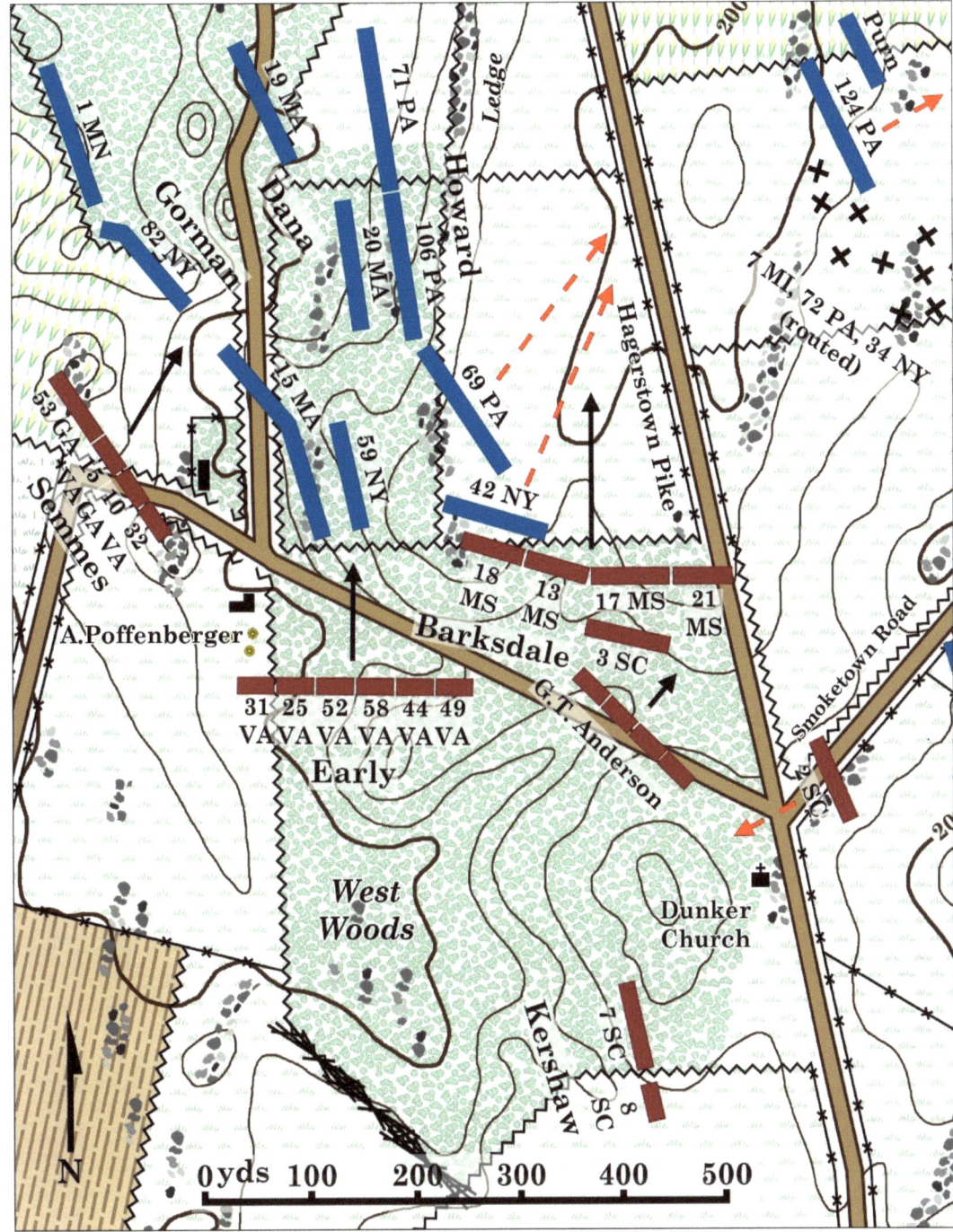

1. 9:35 a.m.
2. Barksdale charges out of the West Woods, driving away the 42nd New York, taking the 69th Pennsylvania in flank, and routing them.[16]
3. Early's Brigade charges into the flank of the 15th Massachusetts and 59th New York.
4. Semmes Brigade charges Gorman.
5. Kershaw's Brigade is fragmented. The 3rd South Carolina follows Barksdale. The 2nd South Carolina advances out of the woods at the Smoketown Road, but is driven back by Tyndale. The rest of the brigade stays in the woods.[17]
6. Routing units overwhelm the green 124th Pennsylvania and carry it from the field.[18]

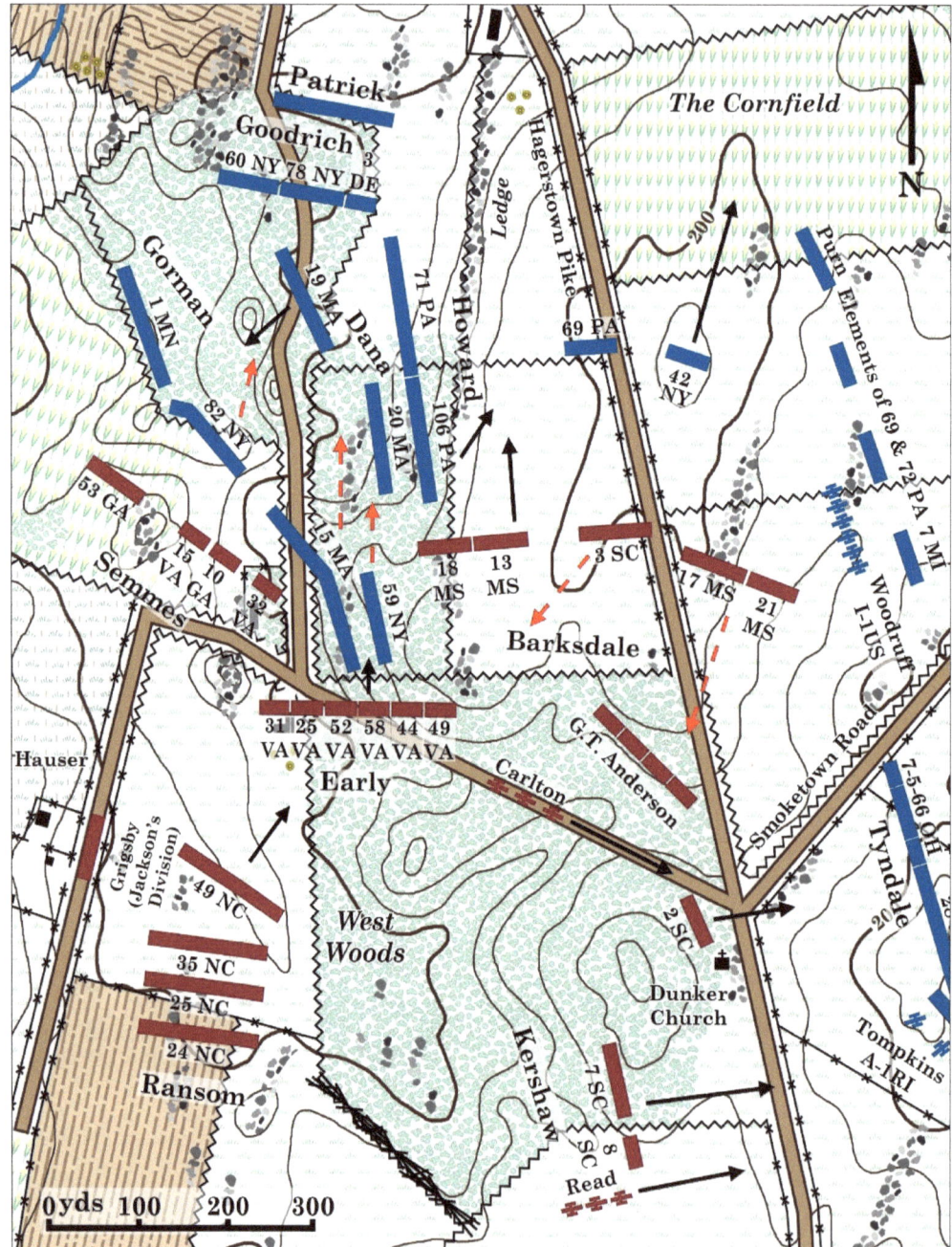

1. 9:40 a.m.
2. Most of the 69th Pennsylvania rallies at the fence and Howard attempts to pull his brigade back and face south to stop Barksdale.
3. Caught by Semmes in front and Early on their flank, Gorman's 15th Massachusetts, 82nd New York, and Dana's 59th New York break and flee north.[19]
4. The left of Barkdale's Brigade and Kershaw's 3rd South Carolina crest the ridge at the Hagerstown Turnpike and come under fire from Battery I, 1st United States and the Union regiments rallying in the field. This fire forces them back to the West Woods.[20]
5. Kershaw orders the 7th and 8th South Carolina forward across the turnpike. The 2nd South Carolina quickly rallies and follows them.[21]
6. Ransom's Brigade arrives and begins to deploy on the western edge of the West Woods.

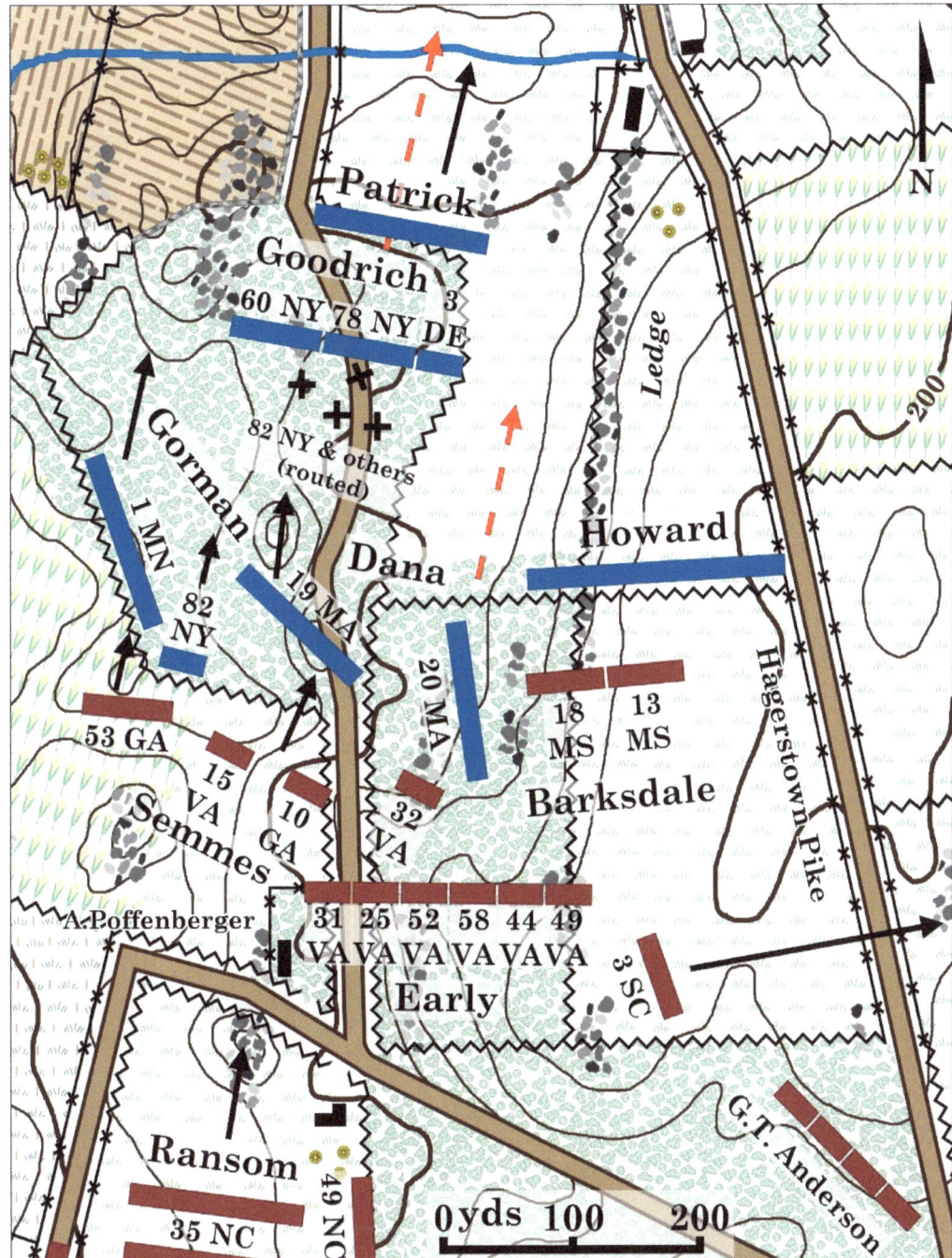

1. 9:45 a.m.
2. Howard reforms along the fence and stops Barksdale's left wing. Howard will then continue his retreat.
3. Surrounded on three sides, the 20th Massachusetts retreats.[22]
4. The 3rd South Carolina rapidly reforms after falling back and moves to the east to rejoin Kershaw's Brigade.
5. Under pressure from Semmes and Early, the 1st Minnesota, 19th Massachusetts, and part of the 82nd New York fall back.[23]
6. Overwhelmed by a fleeing mob, Goodrich's brigade is caught in the rout and leaves.
7. Patrick's brigade falls back in an orderly fashion.
8. Ransom's Brigade continues to move north and deploy.

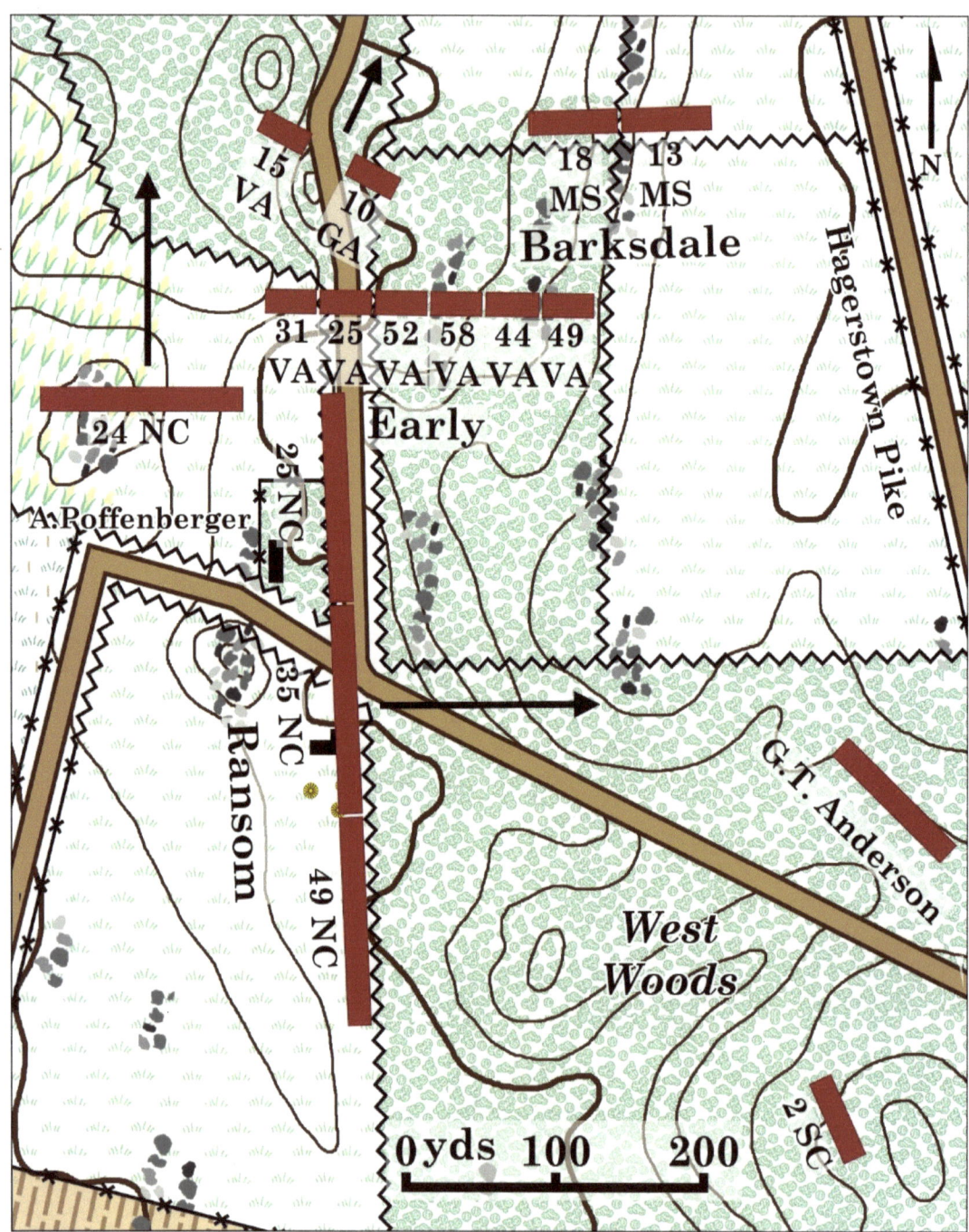

1. 9:50 a.m.
2. Ransom finished deploying and advances to the east.
3. The 24th North Carolina of Ransom's Brigade continues north, following Semmes' Brigade as it pursues Sedgwick's division.
4. Early wheels to the right and takes up a position facing the open fields and the Hagerstown Turnpike.
5. The left wing of Barksdale's Brigade will fall back and reform behind Early's new line.

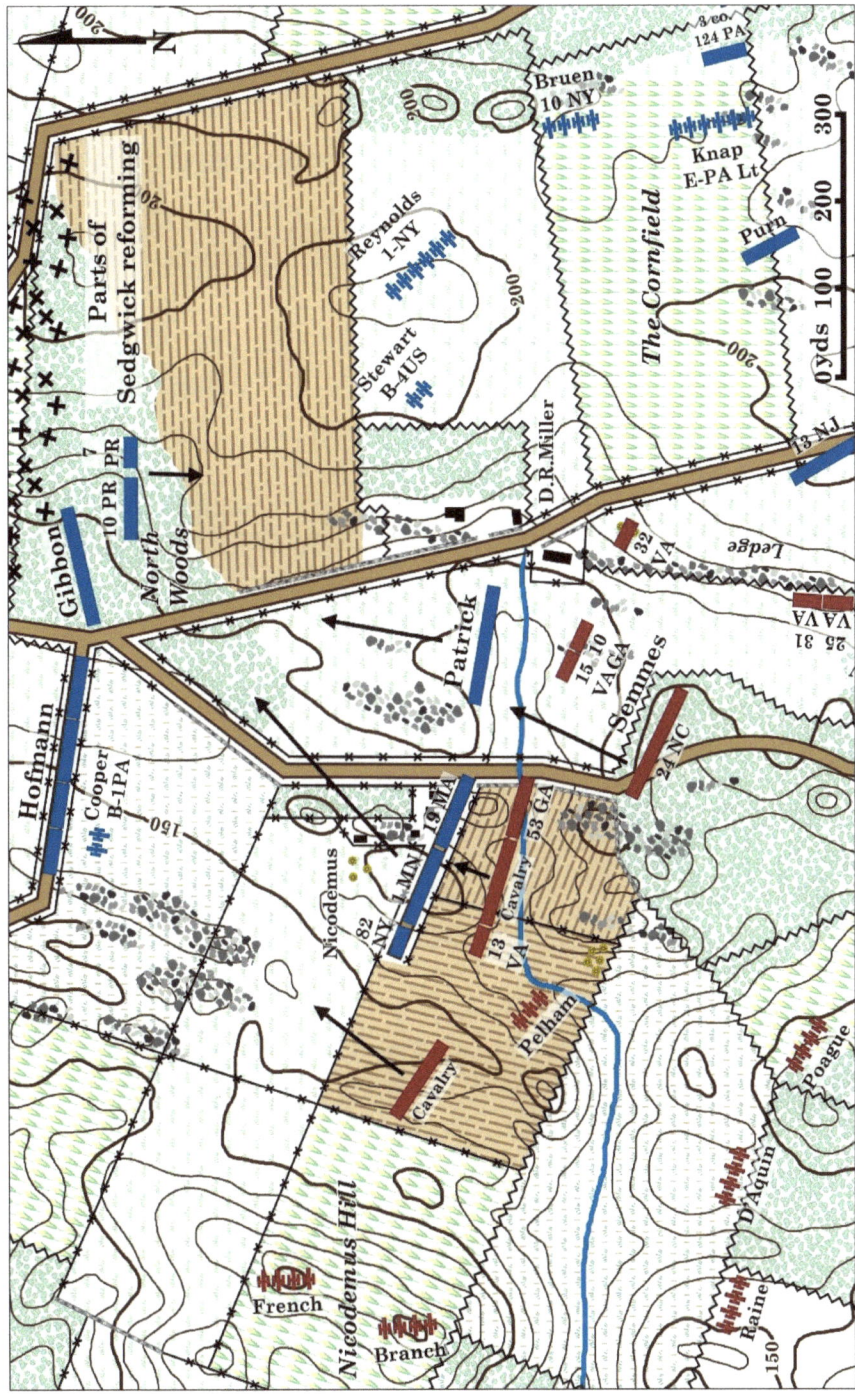

1. 10:00 a.m.
2. Sedgwick's division reforms in the North Woods.
3. Hofmann's brigade moves forward to the lane opposite the North Woods. The 7th and 10th Pennsylvania Reserves move forward to the edge of the North Woods.[24]
4. The 82nd New York, 1st Minnesota, and 19th Massachusetts make a stand in the Nicodemus farm lane to slow down the oncoming Confederates.[25]
5. Patrick's brigade forms on their left.
6. After trading several volleys with the Confederates, both formations fall back in good order.
7. Semmes, the 24th North Carolina, and some of Stuart's cavalry pursue.

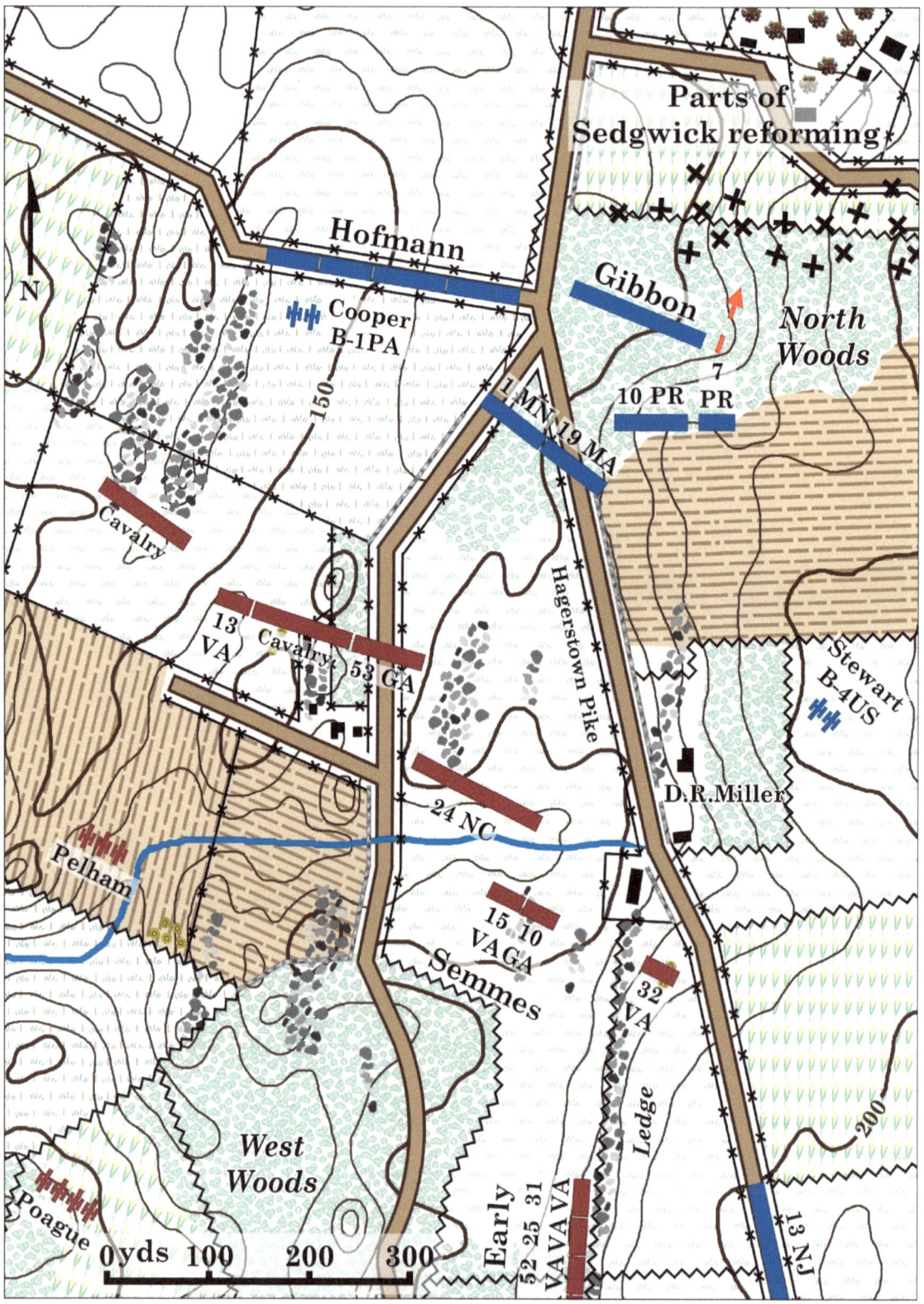

1. 10:15 a.m.
2. The 1st Minnesota and 19th Massachusetts reform in the southwest corner of the North Woods.[26]
3. Fire from Hofmann and these two regiments, help from Stewart's section of Battery B, 4th United States, and exhaustion halt the Confederate forward surge.

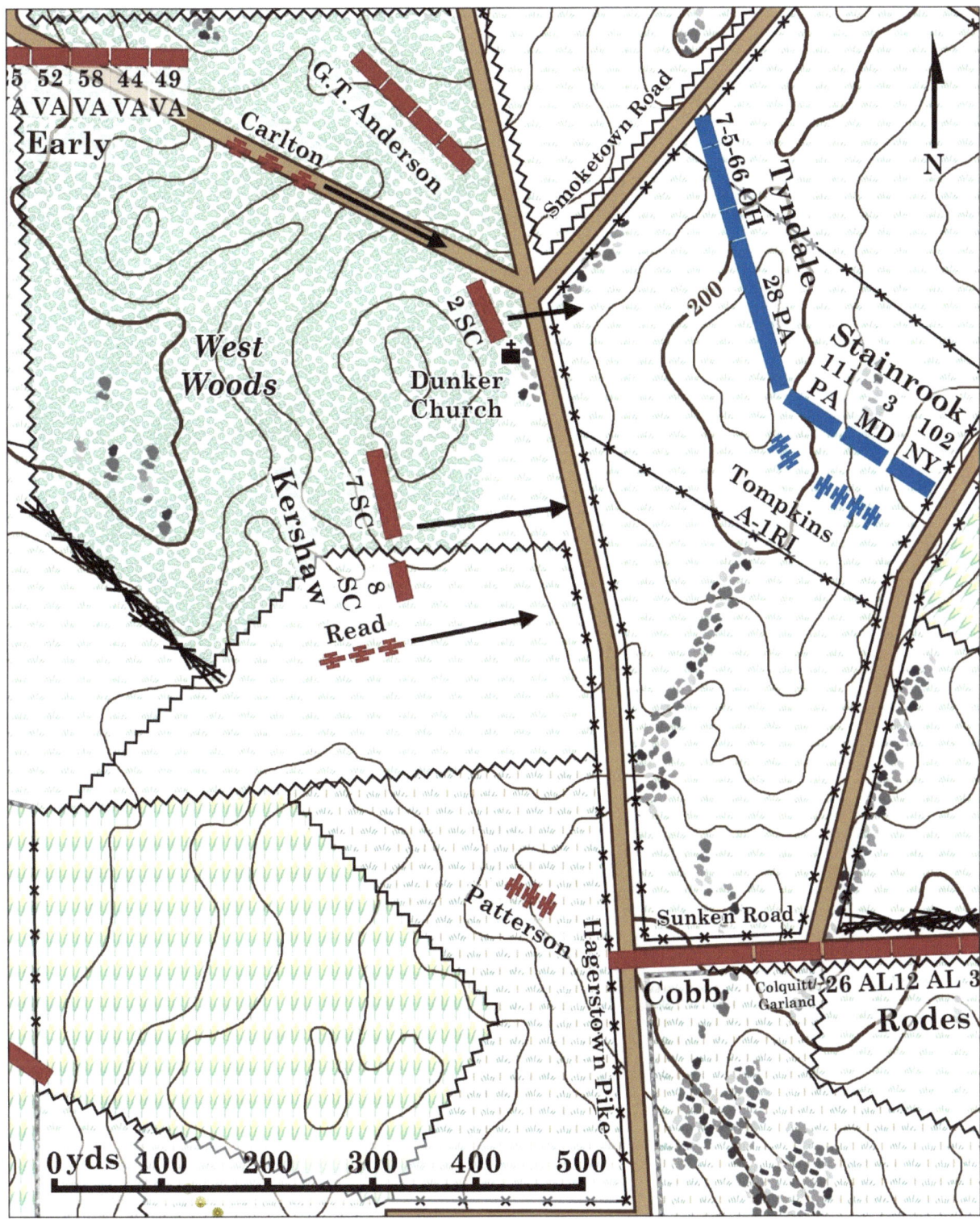

1. 9:40 a.m.
2. Kershaw orders the 7th and 8th South Carolina forward to help the 2nd.[27]
3. Read and Carlton's Batteries move forward to support.[28]

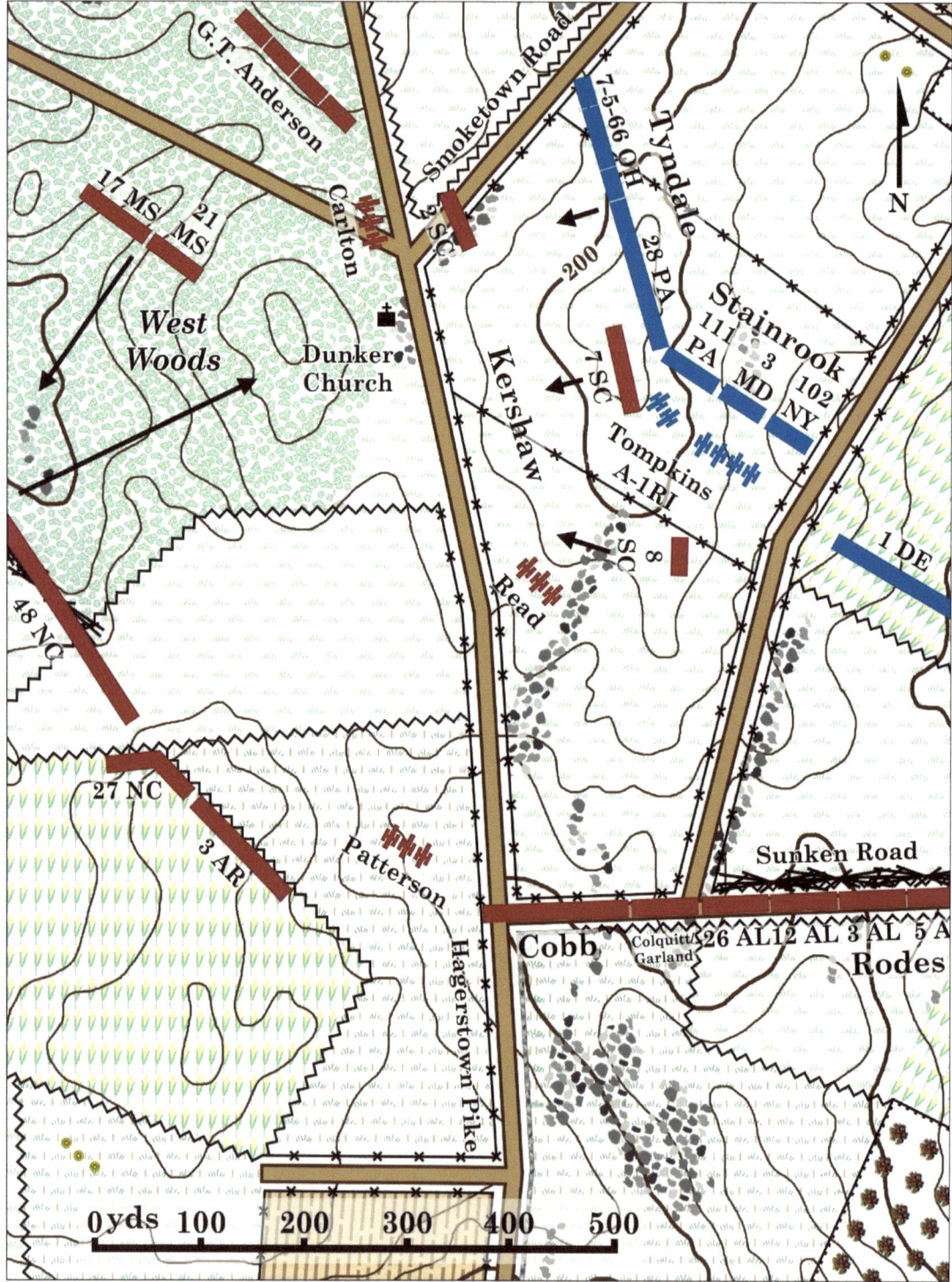

1. 9:45 a.m.
2. The 7th and 8th South Carolina crest the plateau and engage Stainrook and Battery A, 1st Rhode Island.[29]
3. The 8th South Carolina fires into the flank of the 1st Delaware as it approaches the Sunken Road.[30]
4. Tyndale charges to the crest and fire from the two Union brigades drives the South Carolinians back.[31]
5. The three left regiments from Manning's recently arrived brigade march through the West Woods to help Kershaw. The 27th North Carolina and 3rd Arkansas remain behind along a cornfield fence.

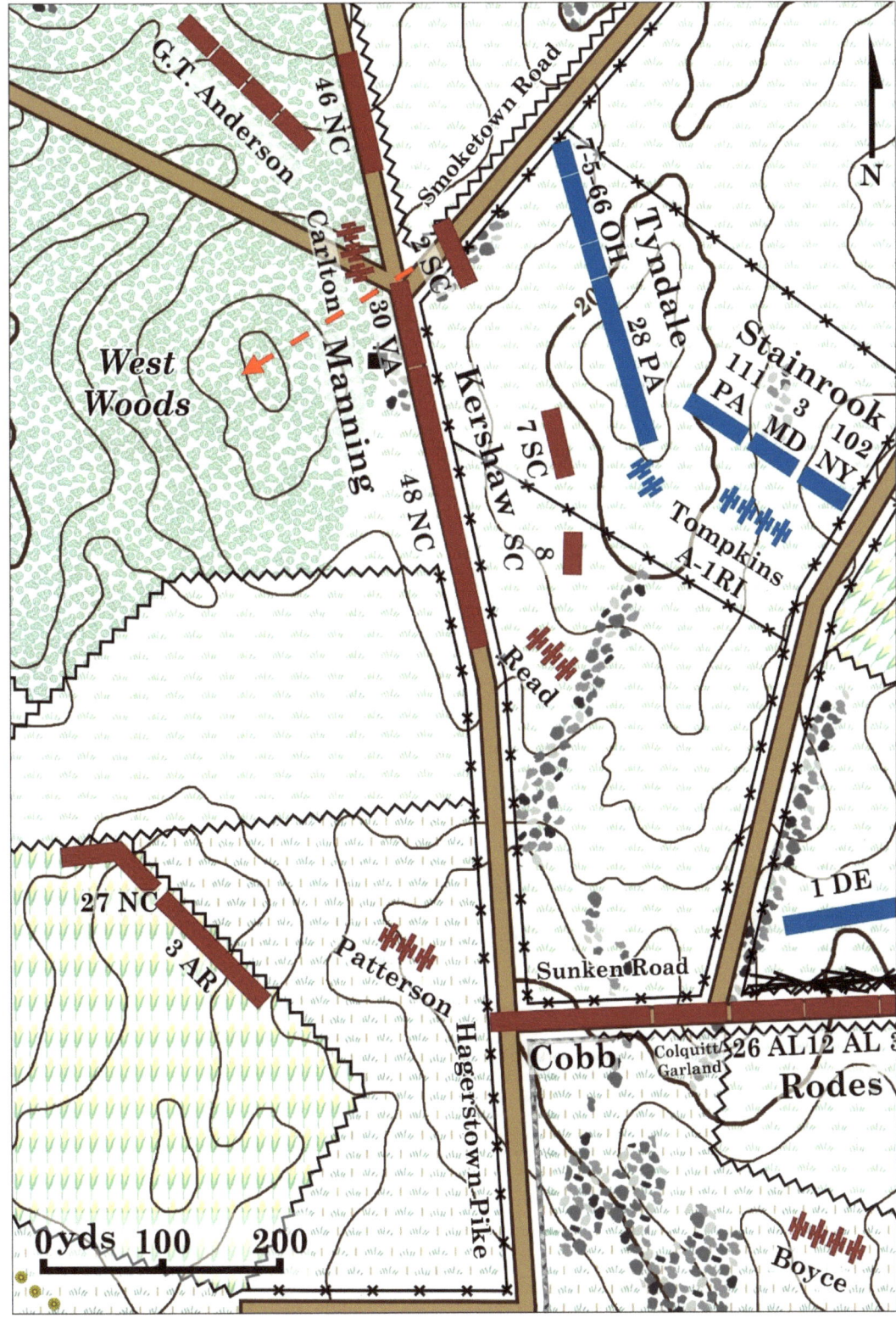

1. 9:50 a.m.
2. Having held out the longest, the 2nd South Carolina retreats back into the West Woods.[32]
3. Manning's Brigade arrives and takes cover along the fences of the Hagerstown Turnpike. They engage Tyndale and the right section of Battery A, 1st Rhode Island.[33]

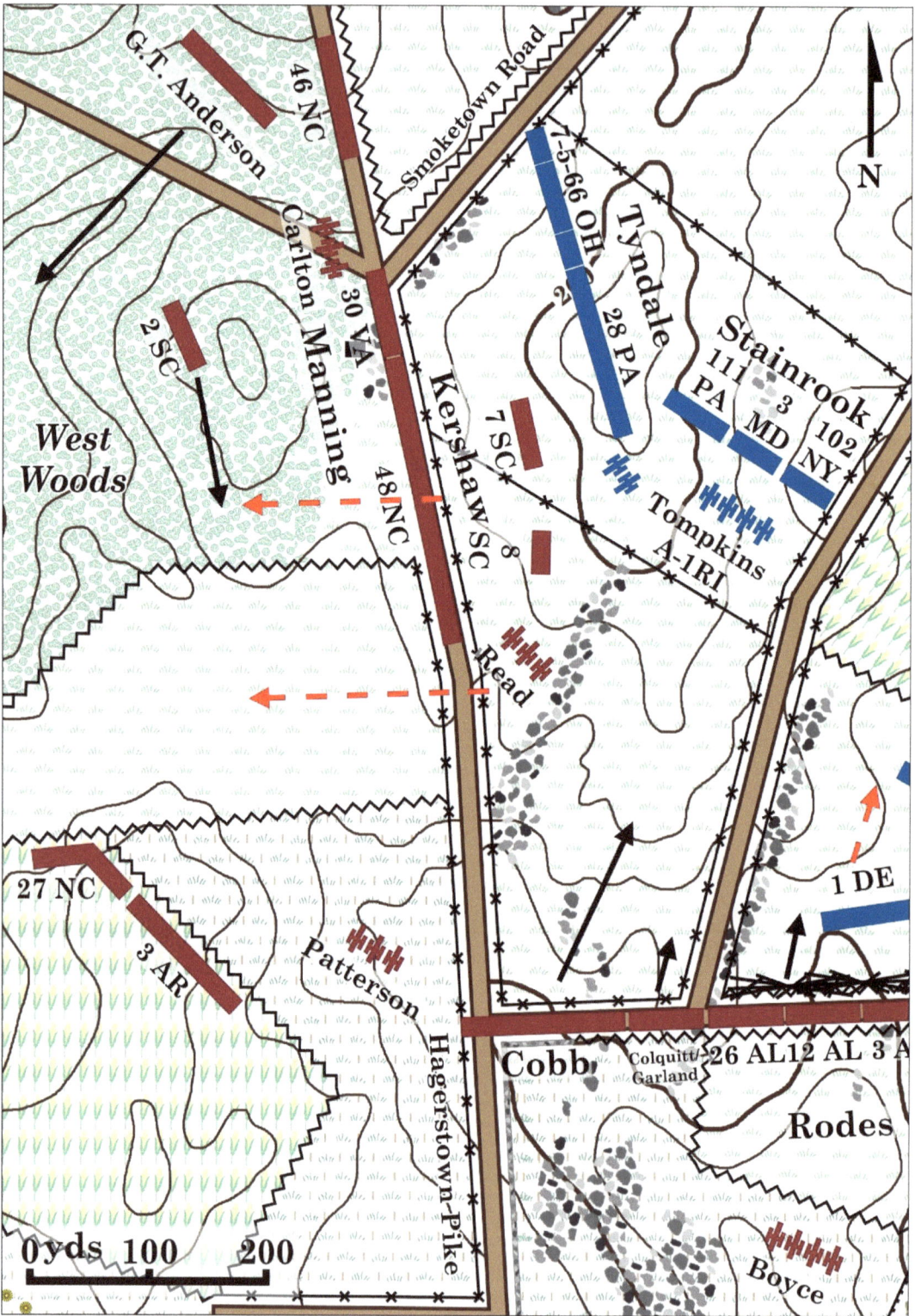

1. 9:55 a.m.
2. The 7th and 8th South Carolina, caught between the lines, retreat and get out of the way.[34]
3. Read's Battery falls back.[35]
4. Cobb's Brigade, the remnants of Colquitt and Garland, and the left of Rodes' Brigade advance from the Sunken Road and push back the 1st Delaware.[36]

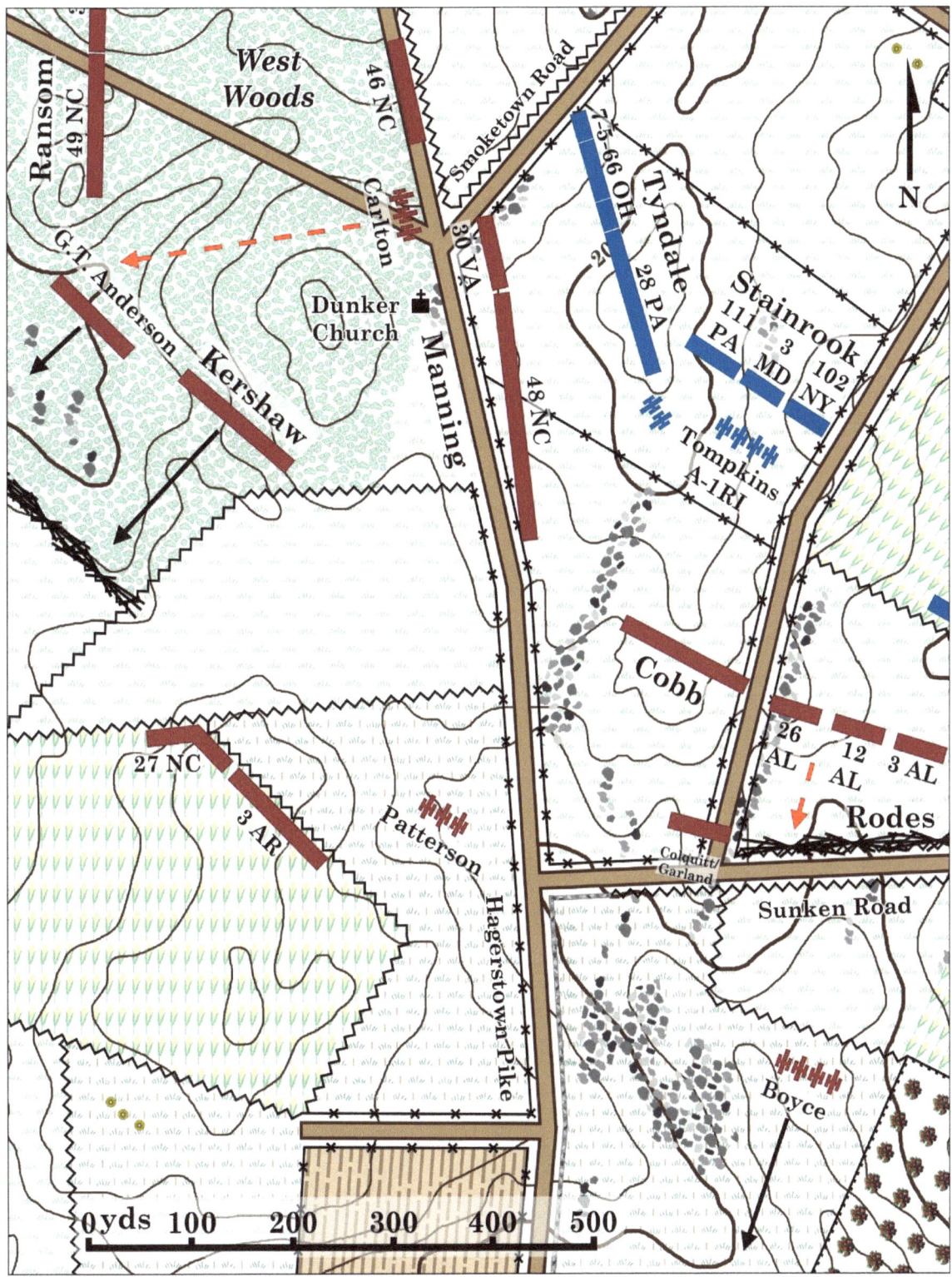

1. 10:00 a.m.
2. Manning advances east of the Hagerstown Turnpike.[37]
3. Carlton's Battery withdraws.[38]
4. G. T. Anderson and Kershaw fall back through the West Woods.
5. Rodes returns to the Sunken Road.

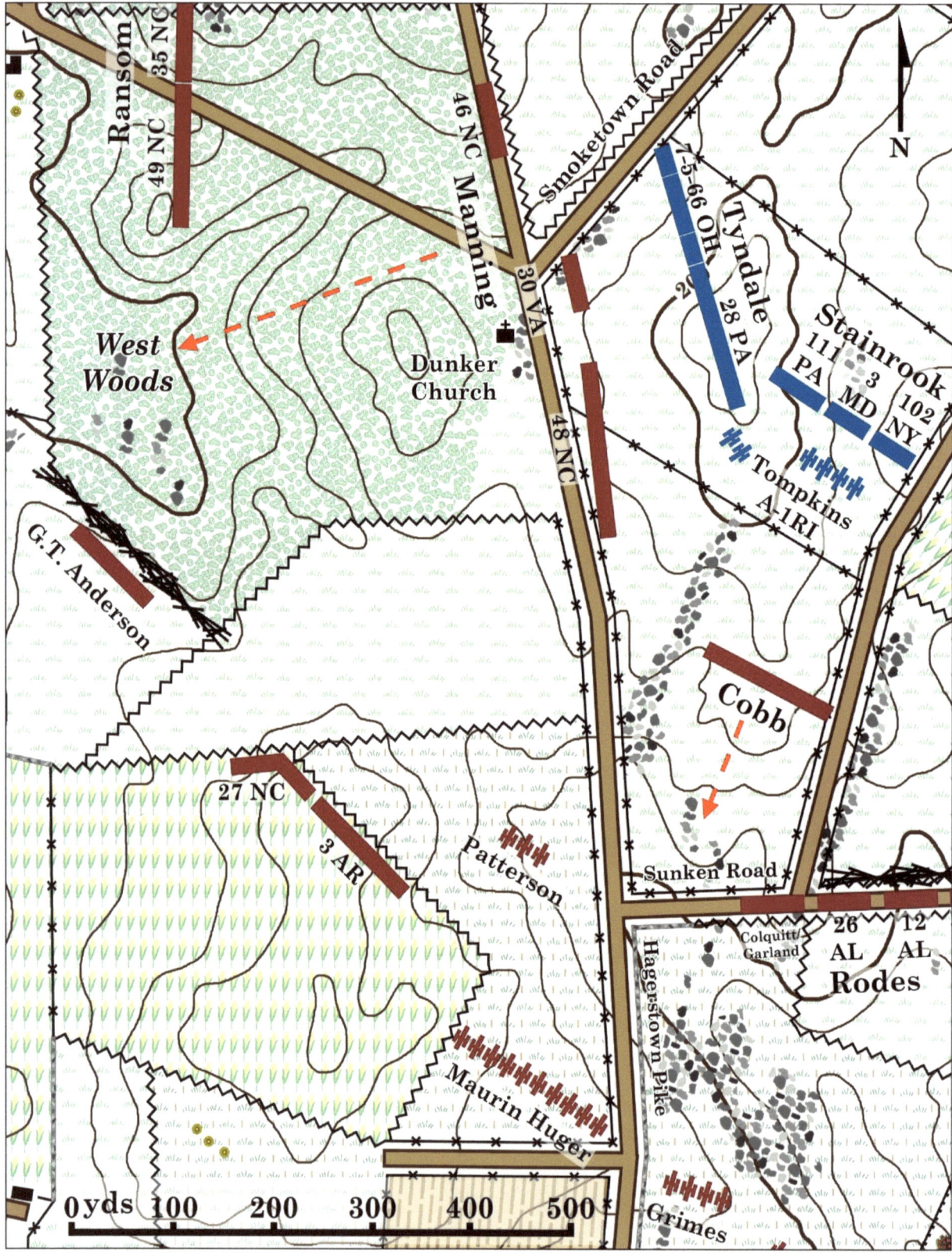

1. 10:10 a.m.
2. Manning falls back into the West Woods.
3. Cobb falls back to the Sunken Road.

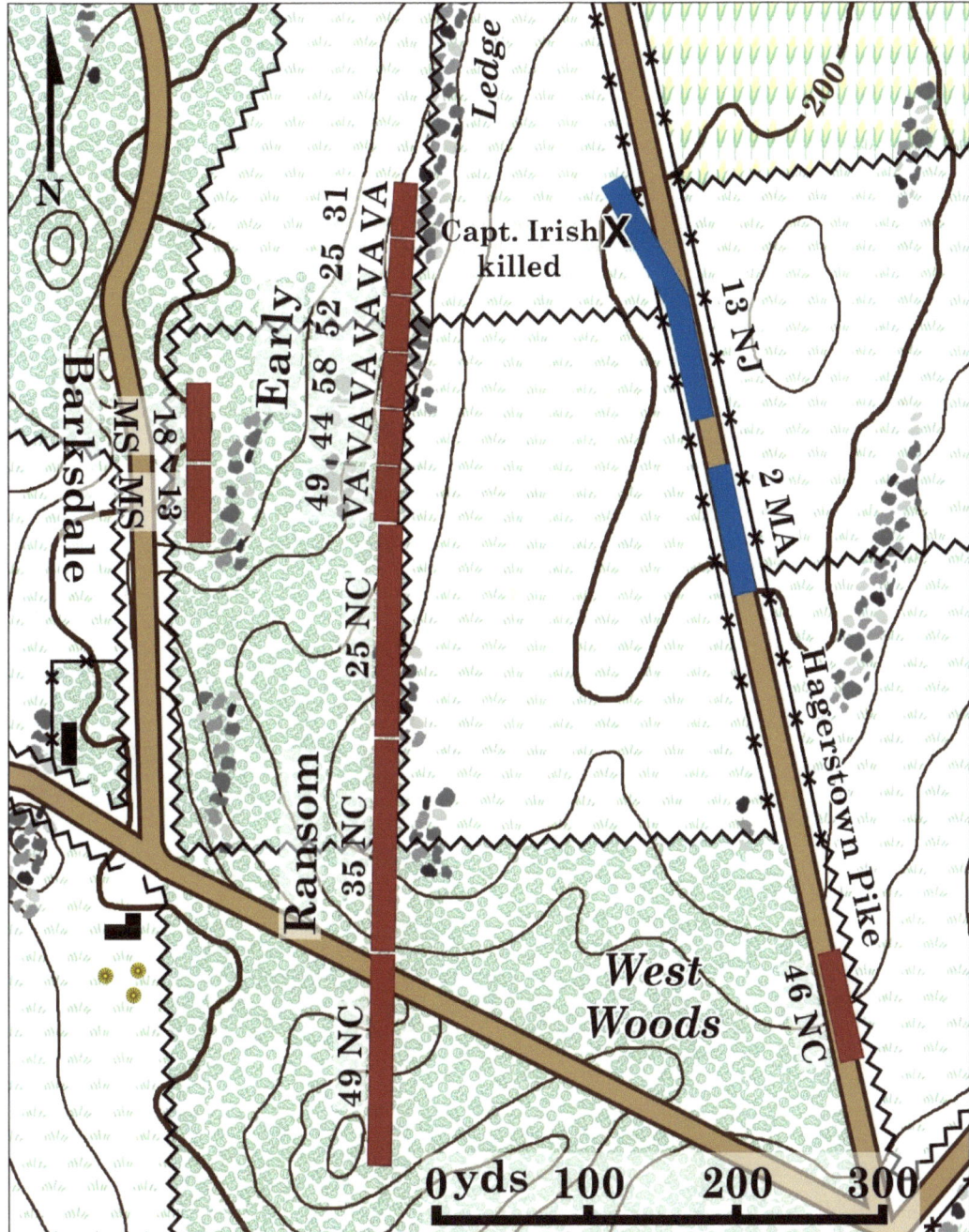

1. 10:00 to 10:10 a.m.
2. The newly arrived 13th New Jersey advances to the Hagerstown Turnpike from the east along with the 2nd Massachusetts. Both belong to Gordon's brigade. They believe that Sedgwick is still in the woods in front and are to support them. The high ground along the turnpike prevents them from seeing the full extent of Sedgwick's rout on the other side.[39]
3. They receive a blistering fire from Early and Ransom along the edge of the West Woods.
4. Captain Hugh C. Irish is killed leading his men across the turnpike fence.[40]
5. Gordon orders them to withdraw since the Confederates are too strongly posted behind the fence and limestone ledge.

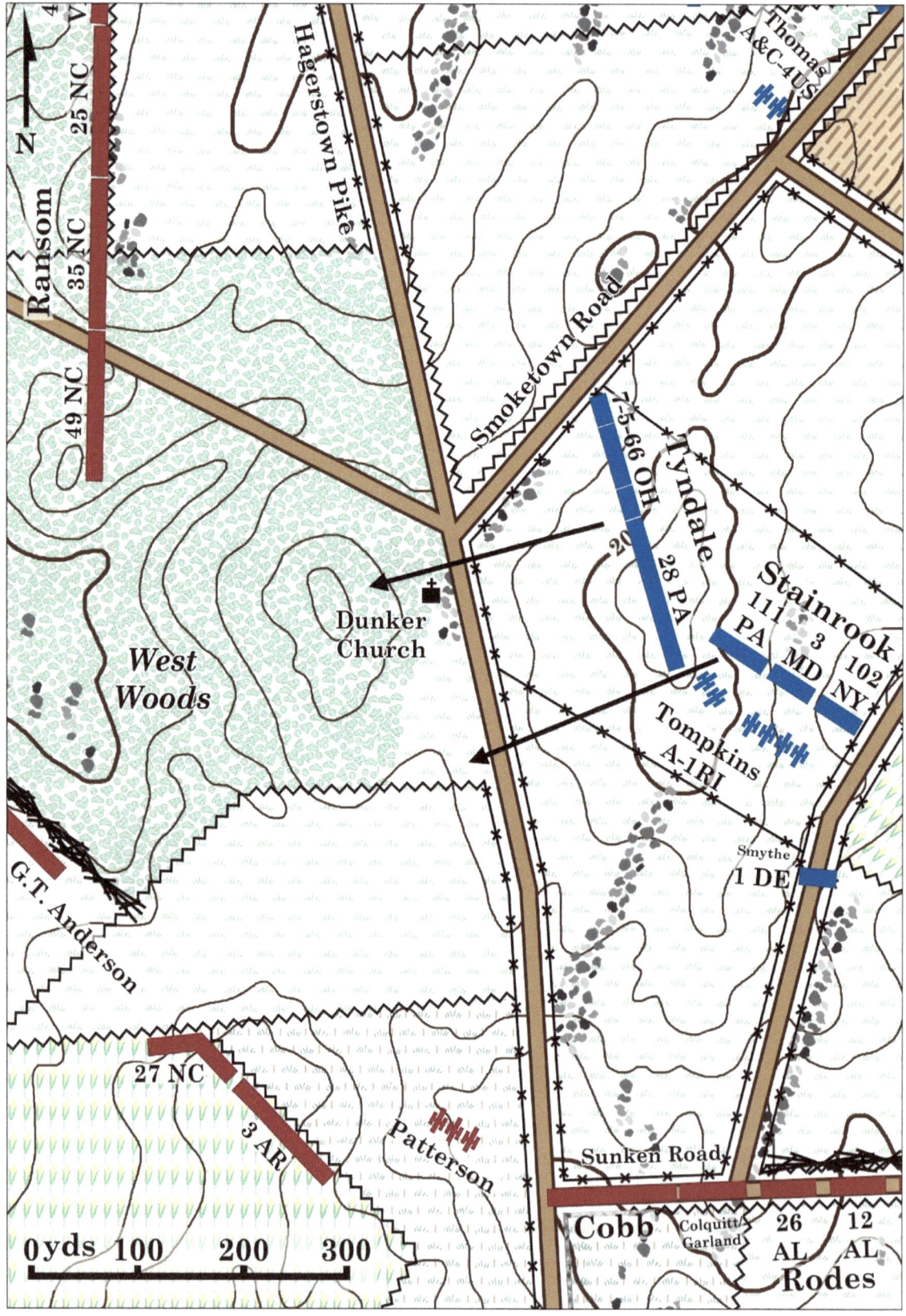

1. 10:20 a.m.
2. Facing no immediate opposition, Greene advances his division into the West Woods.[41]

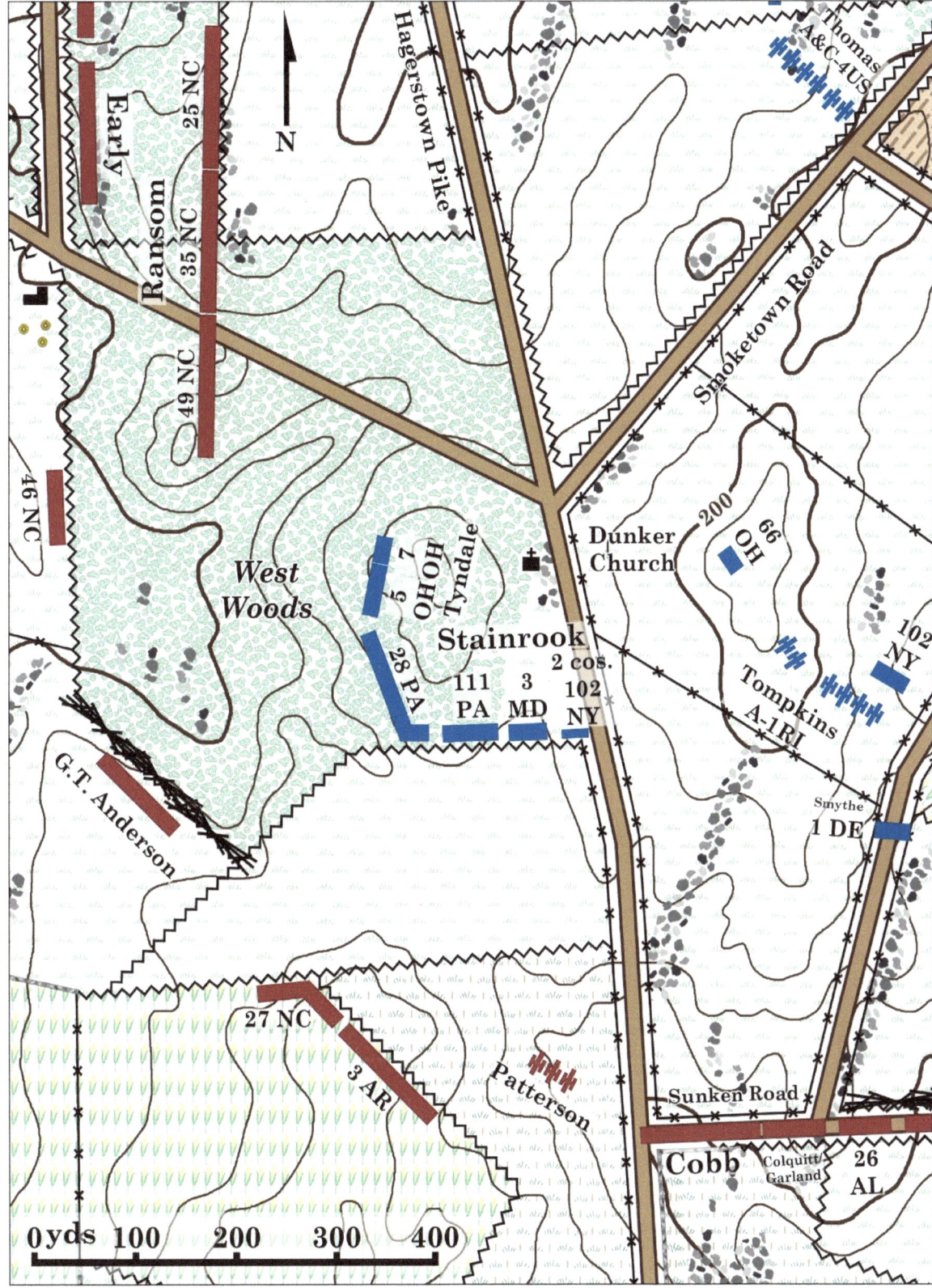

1. 10:30 a.m.
2. Greene sets up a defensive perimeter around the Dunker Church with Tyndale facing west and Stainrook facing south.[42]

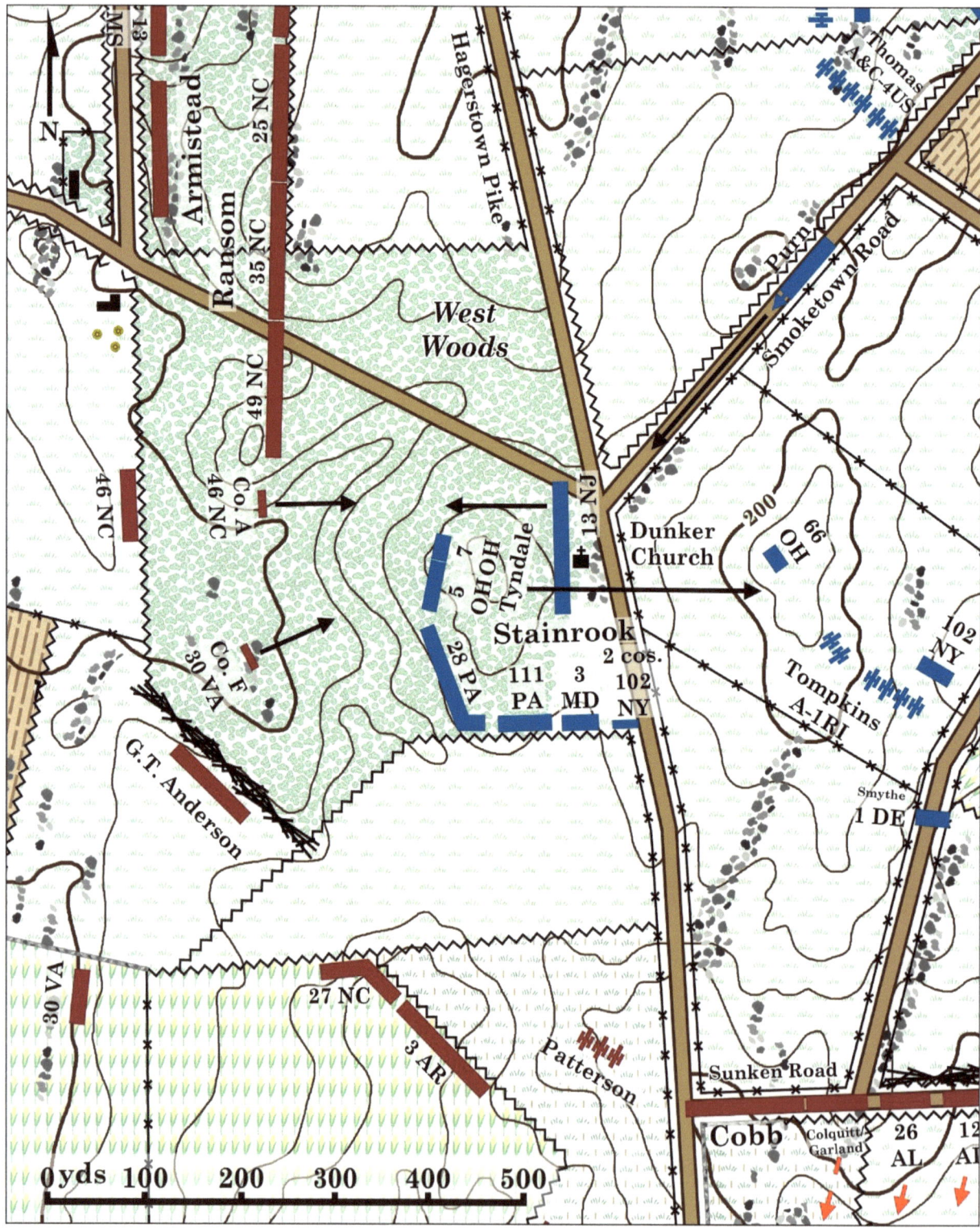

1. 11:00 a.m.
2. Confederates skirmish with Greene.
3. The 5th and 7th Ohio fall back to the plateau. The 13th New Jersey replaces them on the line.[43]
4. The Purnell Legion moves south to reinforce Greene.[44]

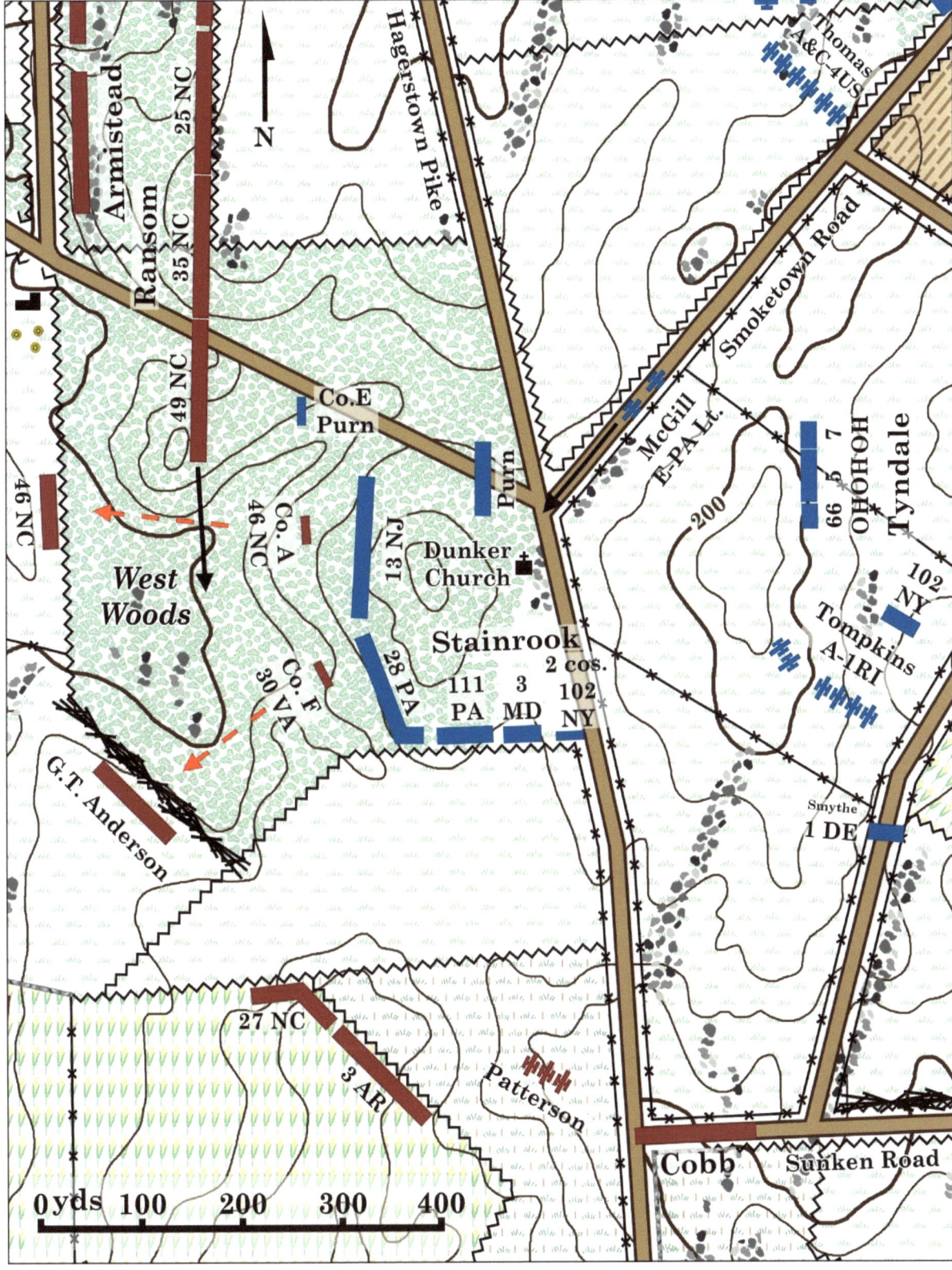

1. 11:10 a.m.
2. The Union line pushes back the Confederate skirmishers.
3. Ransom's Brigade moves south to confront Greene.

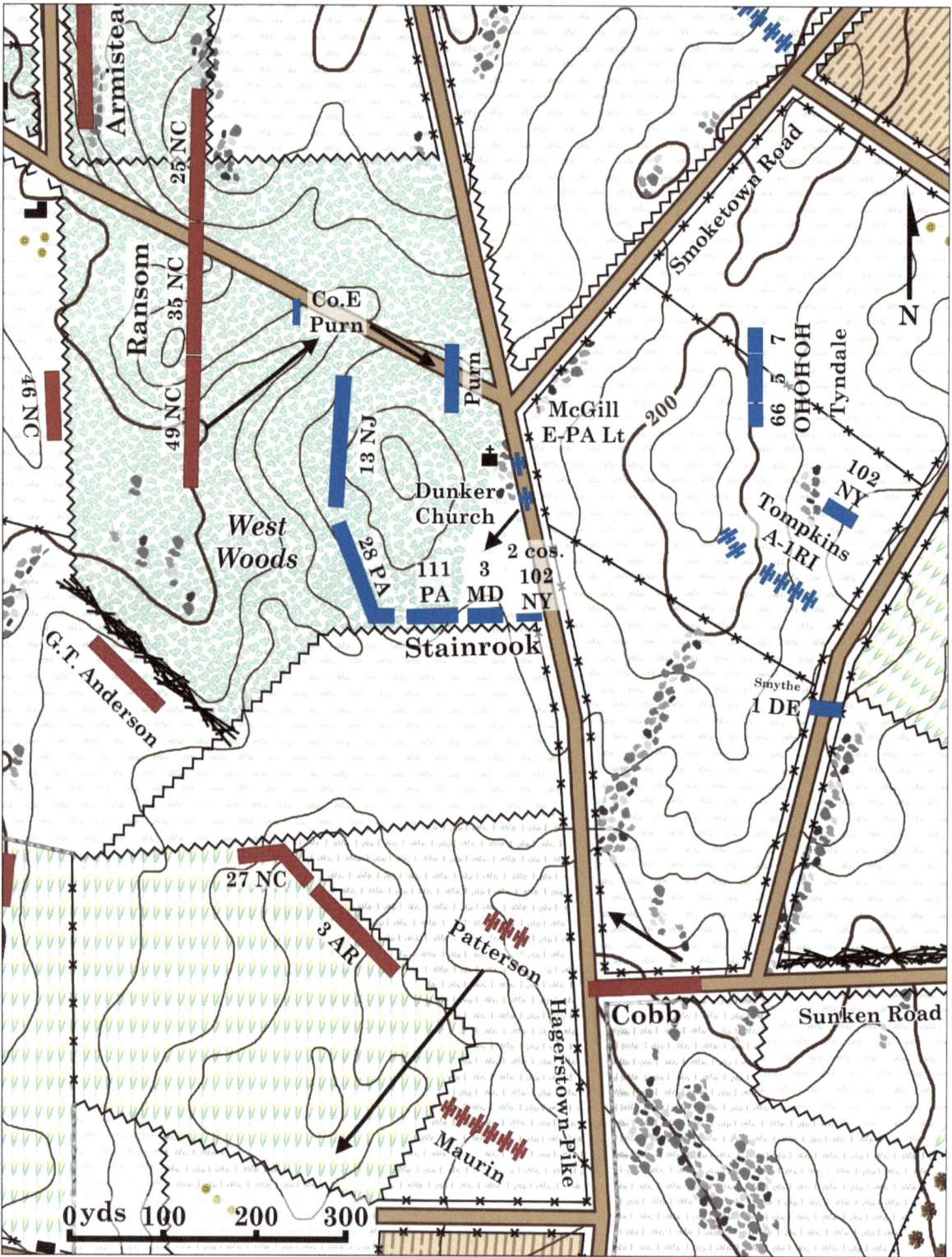

1. 11:15 a.m.
2. Ransom advances up a ravine toward Greene. His intent is to capture McGill's section of Battery E, Pennsylvania Light. He does not realize Greene is stationed around the church.[45]
3. Company E of the Purnell Legion falls back in the face of Ransom's advance.[46]
4. Cobb's Brigade changes front and takes a position in the Hagerstown Turnpike facing east.[47]

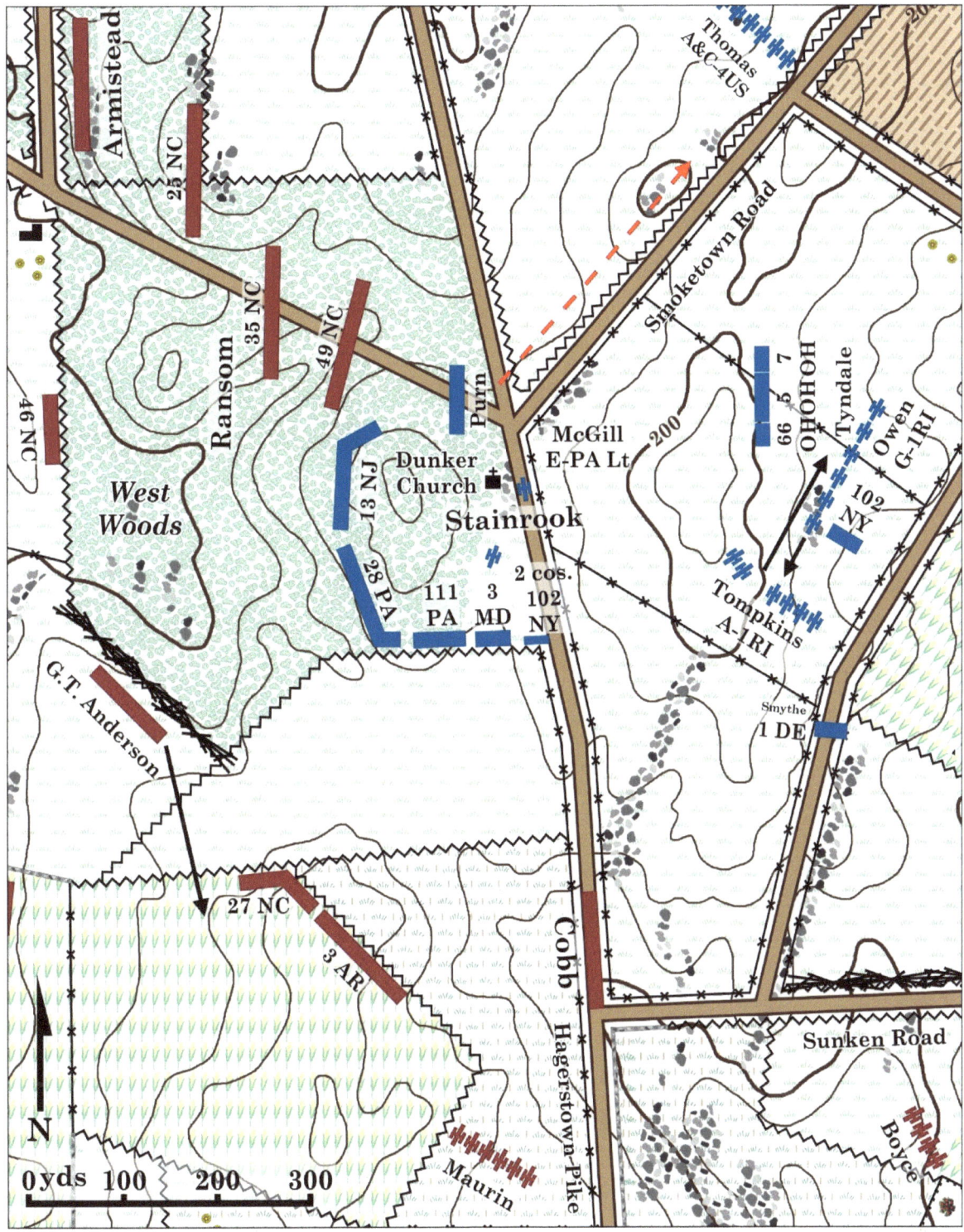

1. 11:30 a.m.
2. The 13th New Jersey, Purnell Legion, and 49th North Carolina are surprised to find an enemy in their front. Officers ride forth to identify each other, but then the infantry begin firing volleys.[48]
3. Purnell falls back.[49]
4. Battery G, 1st Rhode Island replaces Battery A, 1st Rhode Island on the plateau.[50]
5. George. T. Anderson moves his brigade south.

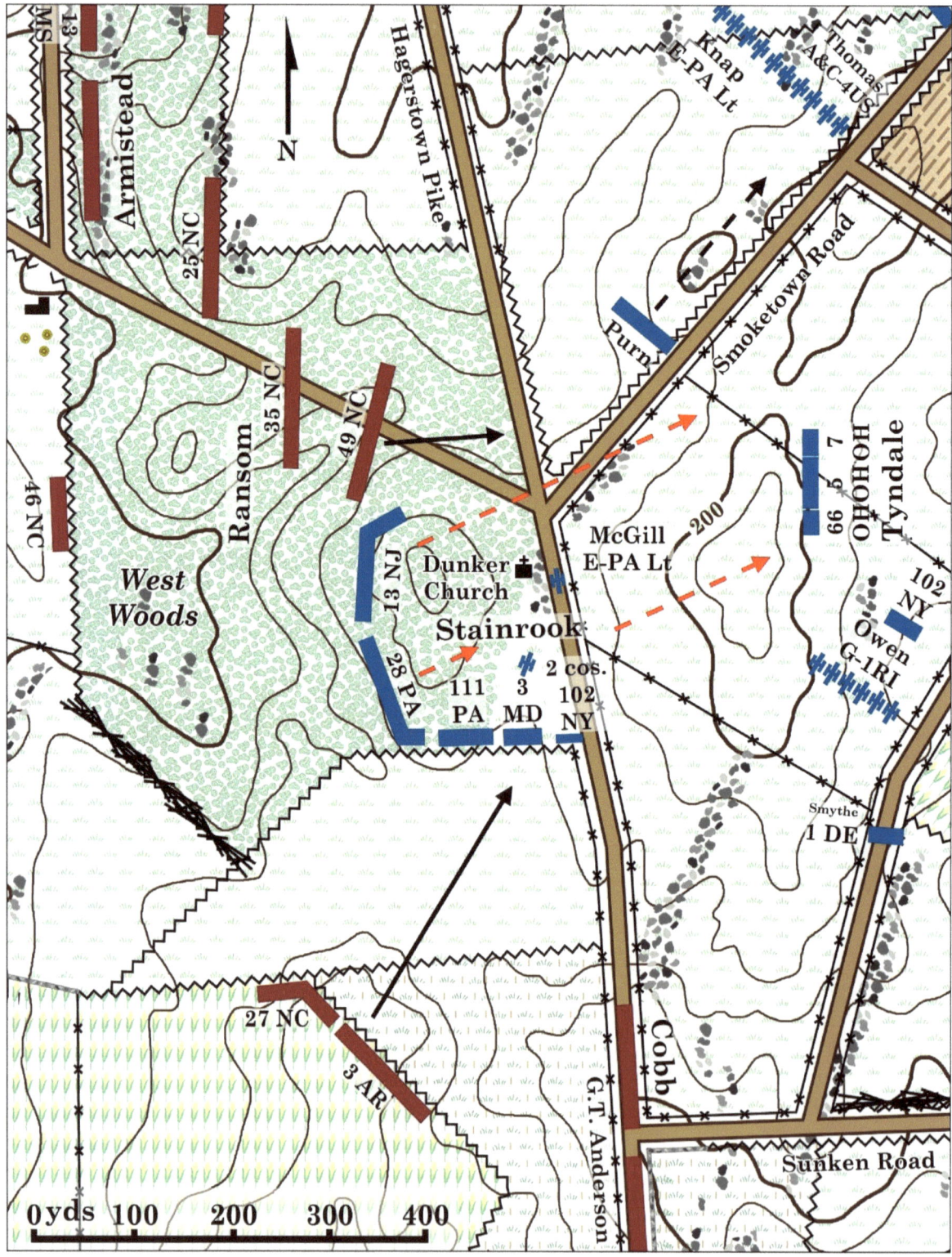

1. 11:45 a.m.
2. The 49th North Carolina charges into and past the flank of the 13th New Jersey.
3. The right regiments of Manning's Brigade, the 27th North Carolina and 3rd Arkansas commanded by the 27th's John R. Cooke, charge Stainrook looking to capture Battery E, Pennsylvania Light.[51]
4. Greene's division retreats out of the West Woods under pressure from both flanks.

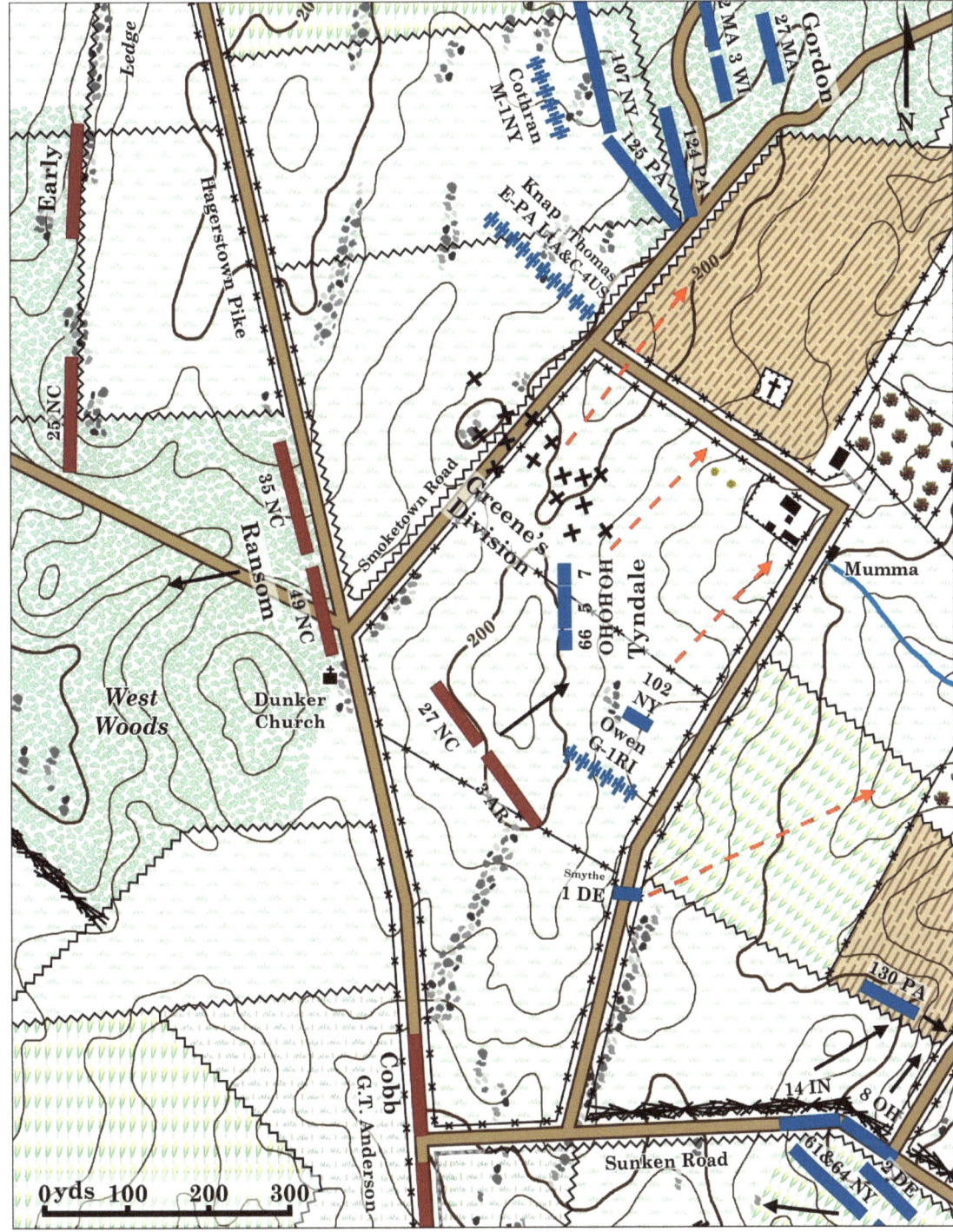

1. 12:00 p.m.
2. Ransom stops at the edge of the West Woods. He is confronted by a line of Union batteries in the field beyond. After a severe shelling, the 35th and 39th North Carolina fall back deeper into the woods.[52]
3. Cooke continues to charge with his two regiments. They reach the crest of the plateau and force the remaining Union infantry and artillery to retreat.
4. After clearing the plateau, Cooke continues his charge to the east into the flank of the mass of Union units on the other side of the Mumma cornfield. See page 85.

The Sunken Road

On the other side of Greene, French moved due south from the East Woods. Here the Confederates under Daniel. H. Hill were waiting for him at a weathered down farm lane called the Sunken Road. It also enjoyed a reverse slope defense, only allowing the attacker to engage the defenders at point blank range. French's three successive brigades mangled themselves against the Sunken Road. Only the arrival of Richardson's division broke the stalemate. They devastated Confederate Major General Richard H. Anderson's division attempting to reinforce the position, and then flanked the length of the Sunken Road. The defenders fell back to the Henry Piper farm.

The Confederates managed to rally quickly, and even sally forth a limited counterattack that delayed the Union advance. Richardson had to commit his last brigade to fend off the rebel advance. More units had to front west to guard against Walker's attack from the West Woods. Once this was achieved several regiments marched south as far as the Piper farm buildings, but they were too exposed and isolated to remain. They fell back, and the Union forces organized into a solid defensive line that stretched from the high ground north of the Sunken Road through the East and North Woods.

The two lines then settled in for the remainder of the day and licked their wounds. Jackson ordered Major General J. E. B. Stuart's cavalry north to find the Union flank and turn it. Instead he found a stout Union defense anchored on high ground backed by an imposing and impenetrable line of artillery. Near the Sunken Road General Richardson was mortally wounded by an exploding shell as the artillery from both sides dueled each other.

The day ended with a desperate and quixotic charge by the 7th Maine into the teeth of the Confederate defenses.

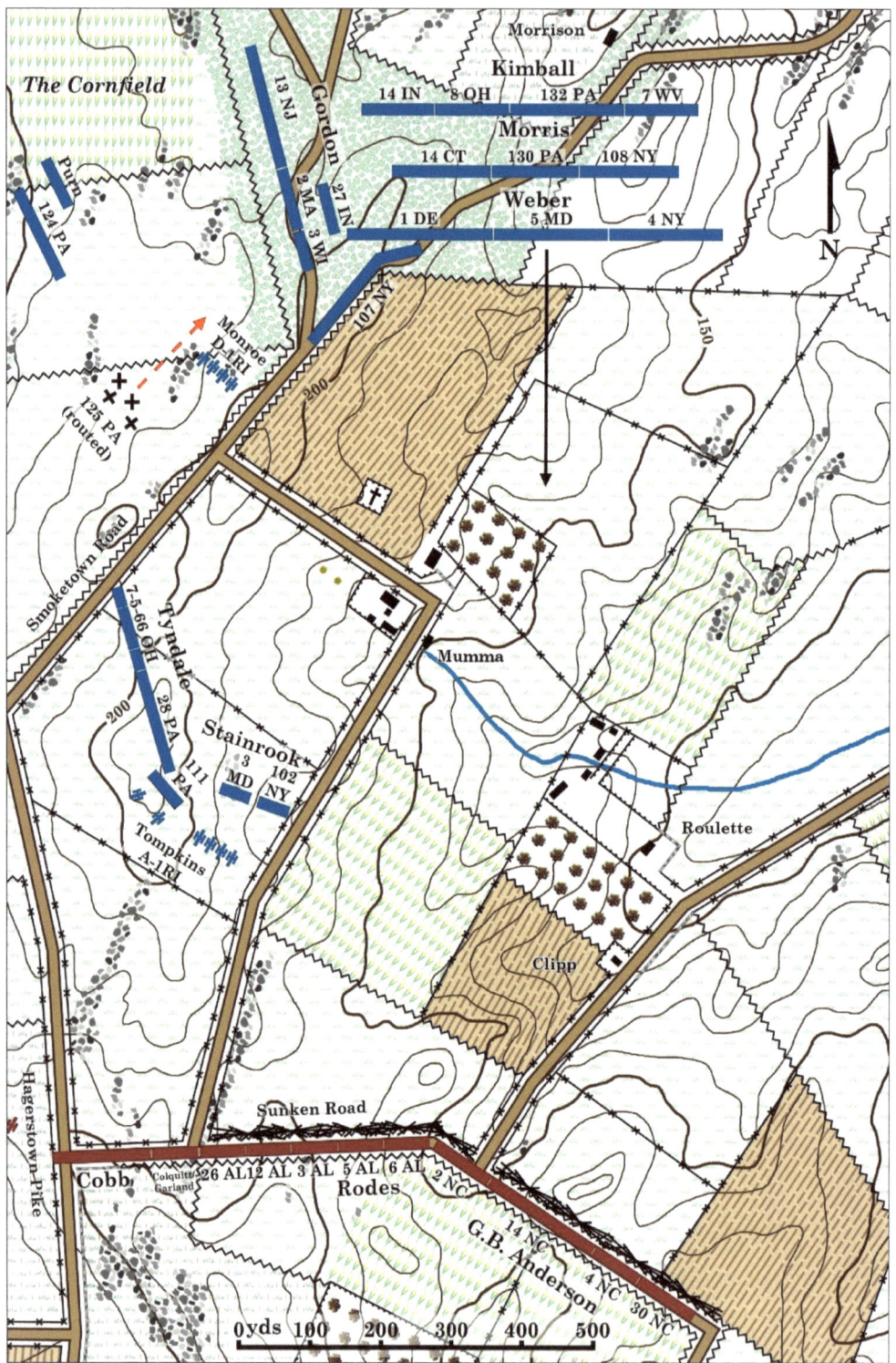

1. 9:30 a.m.
2. French's division leaves the East Woods and marches south. They are to link up with Greene's division on their right, with Sedgwick on Greene's right, and form a solid front to advance on the Confederates. Richardson's division would then form on French's left.[1]
3. Unfortunately, the Twelfth and Second Corps units in the West Woods are already starting to unravel. French will unknowingly be advancing alone.

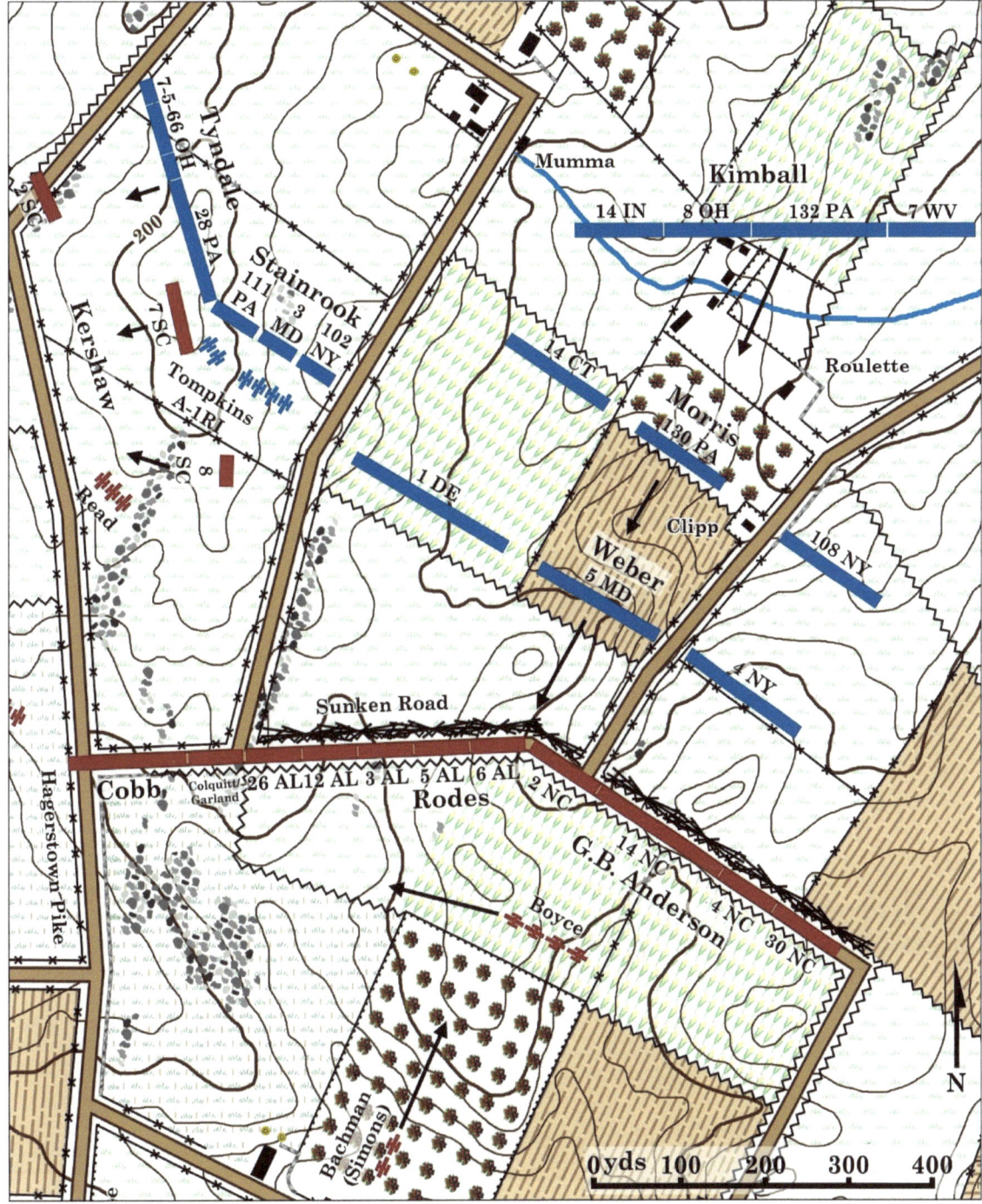

1. 9:45 a.m.
2. French's division marches into battle in the same formation as Sedgwick; one brigade behind the other in a column. However, the interval is far longer and the ground much more open.
3. The lead brigade under Weber advanced on the Sunken Road. The road is in a reverse slope position. The attacking Union forces cannot see the Confederate defenders until they crest the high ground only yards in front of the road, resulting in point blank rifle fire. The exception is the center and left of Rodes' Brigade, where the ground is gentler and has a longer line of sight toward the Mumma farm.
4. The 8th South Carolina briefly fires into the right flank of the 1st Delaware, but does not stop the Union regiment. Tyndale's counterattack and fire from Battery A, 1st Rhode Island drives them off.[2]

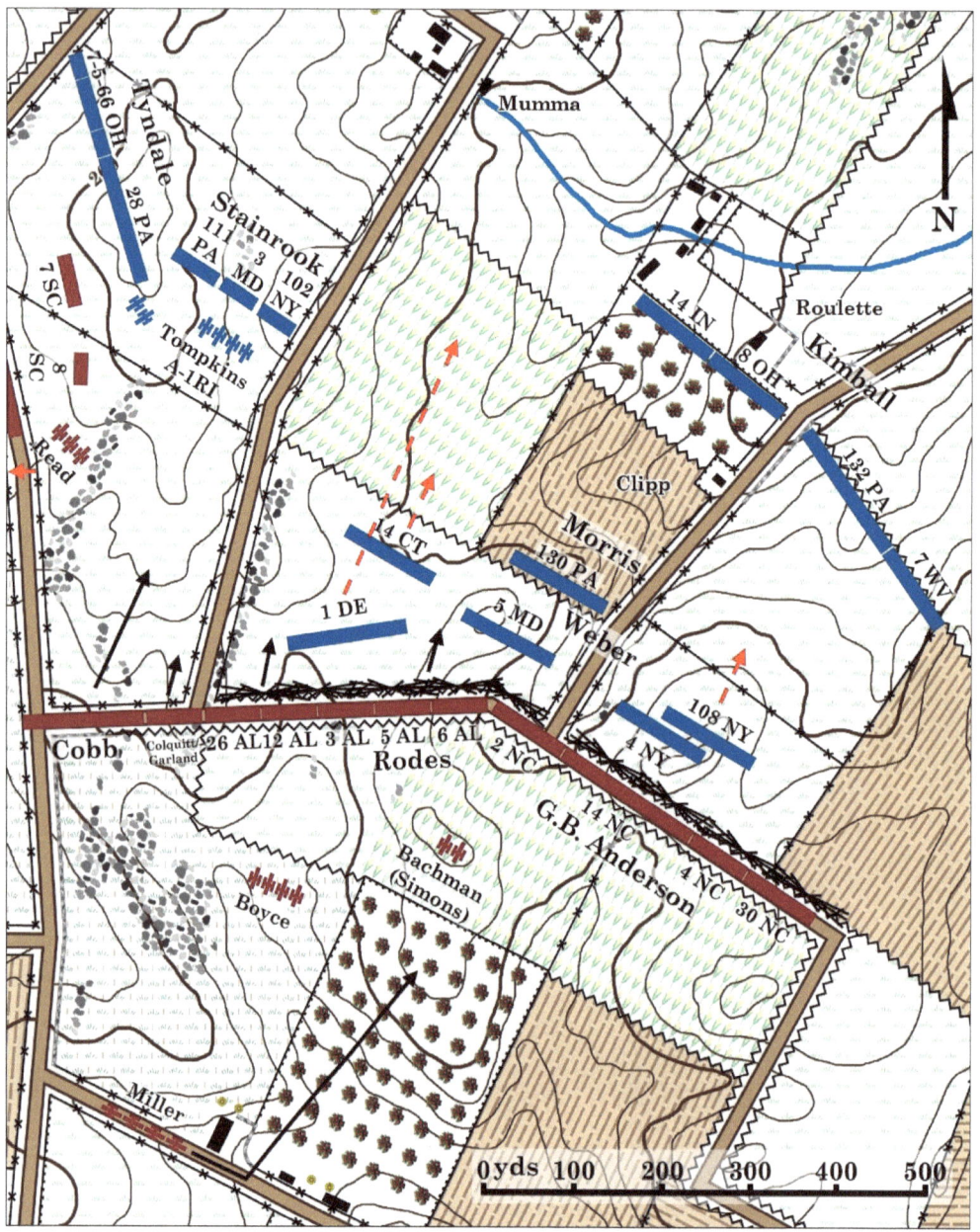

1. 9:55 a.m.
2. Weber's attack stalls in front of the Sunken Road.
3. Morris moves his brigade forward and the two lines intermingle, but they begin to lay a volume of fire on the Confederates.
4. The 108th New York receives such a heavy fire that it recoils and falls back to the north slope of the hill.[3]
5. The 14th Connecticut advances out of the Mumma cornfield and receives the same intense fire from Rodes, Cobb, and Garland/Colquitt directed at the 1st Delaware. The 14th fires a volley, striking the Delawareans in the rear. The 1st falls back through the 14th until they reach the north end of the Mumma cornfield. The 14th falls back to the southern fence of the cornfield.[4]
6. Longstreet orders Rodes to charge his brigade out of the Sunken Road, followed by Cobb and Garland/Colquitt.[5]
7. Confederate artillery batteries begin to arrive to bolster the defense.

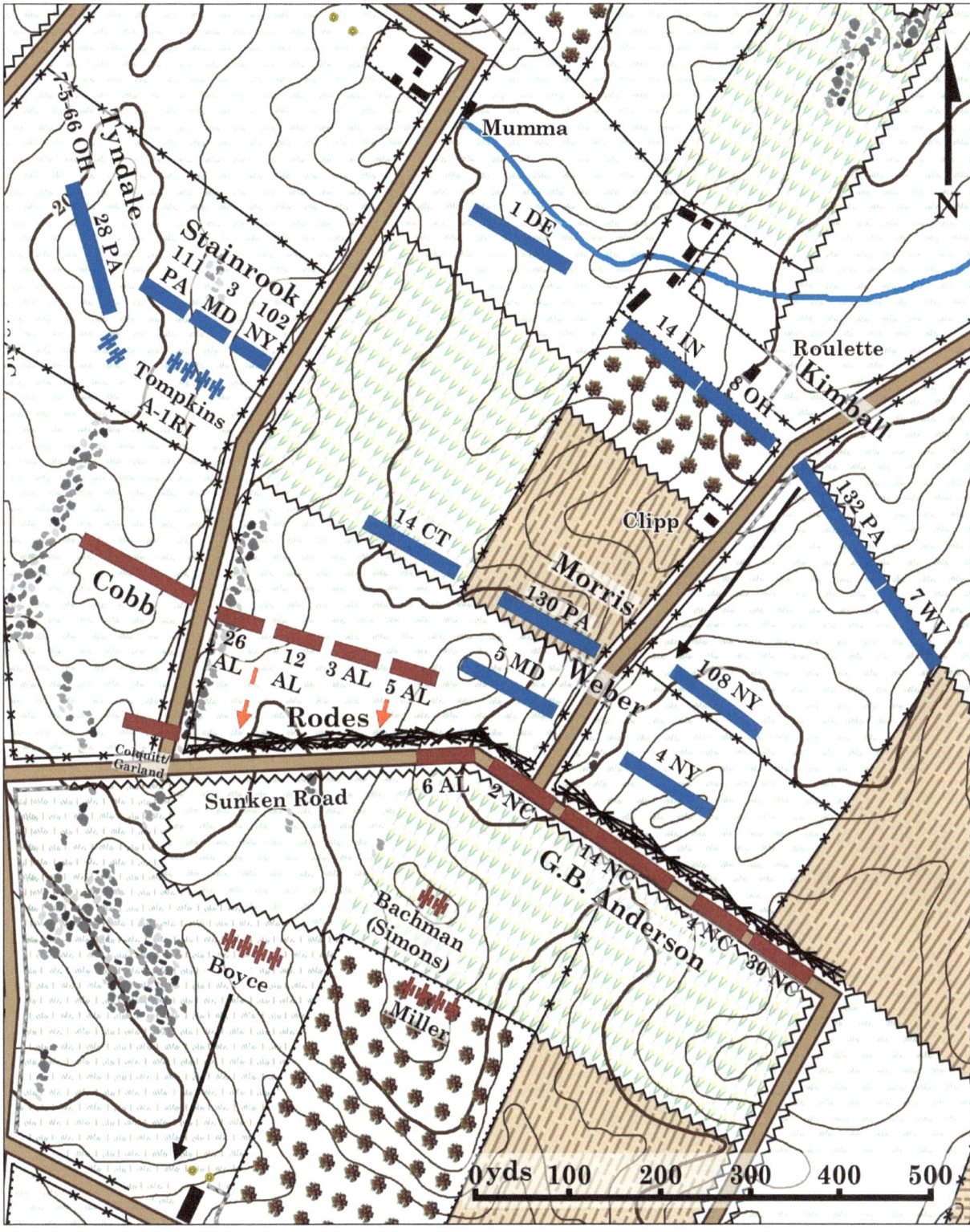

1. 10:00 a.m.
2. The 6th Alabama does not hear Rodes' order to advance, and Garland/Colquitt do not advance as far. Fire from Battery A, 1st Rhode Island savages the brigade. Rodes quickly falls back and rallies in the Sunken Road.[6]
3. Kimball's brigade advances to support Weber and Morris.

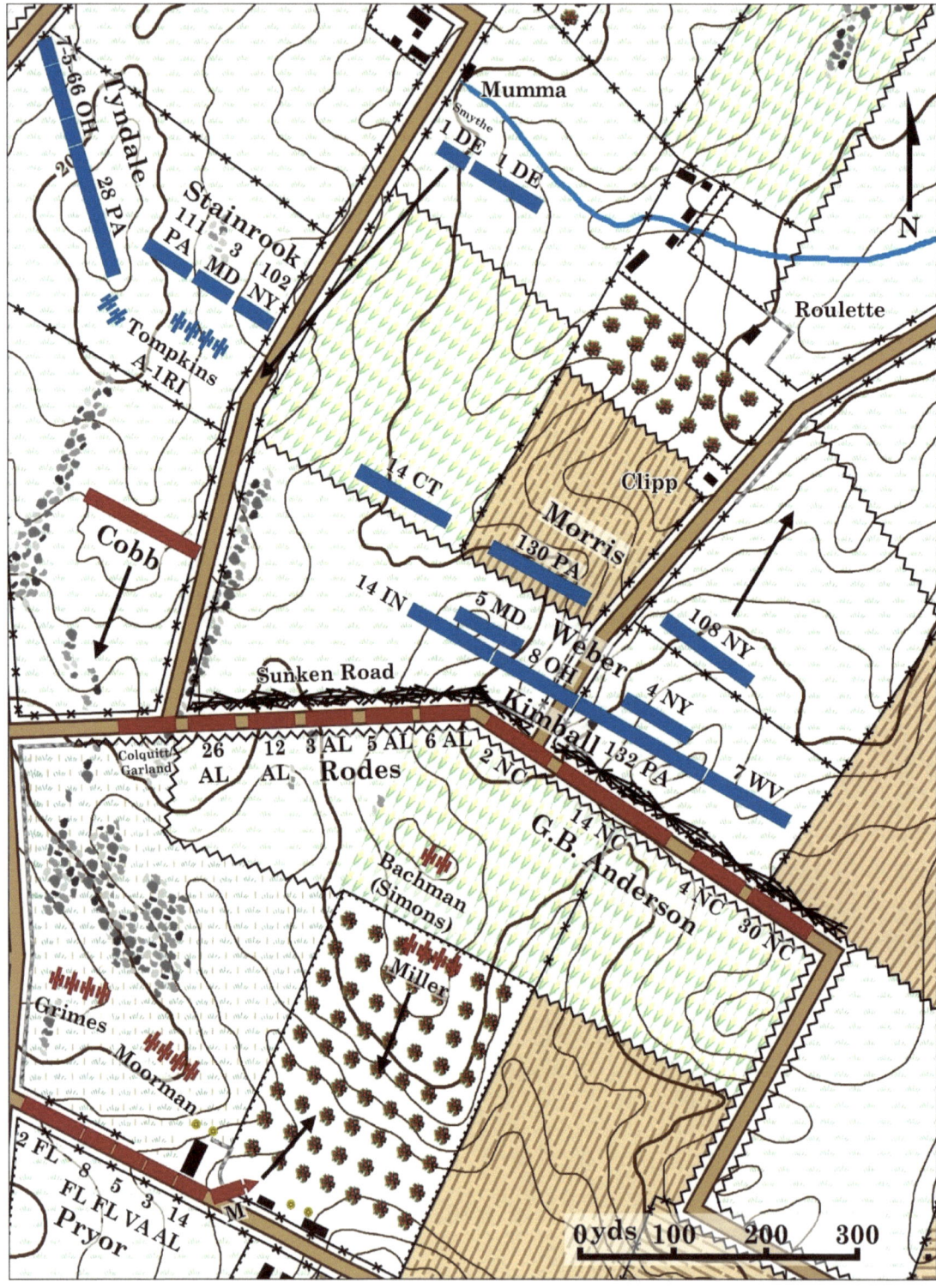

1. 10:15 a.m.
2. Kimball's brigade advances and passes through Morris and Weber. Confederate fire stops them.
3. About 90 men from the 1st Delaware under Major Thomas A. Smyth advance to the rock outcroppings south of the Mumma cornfield and skirmish with the Confederates to the south.[7]
4. Cobb's Brigade falls back from its advanced position.
5. Richard H. Anderson's division begins arriving. Pryor's Brigade leads the advance.

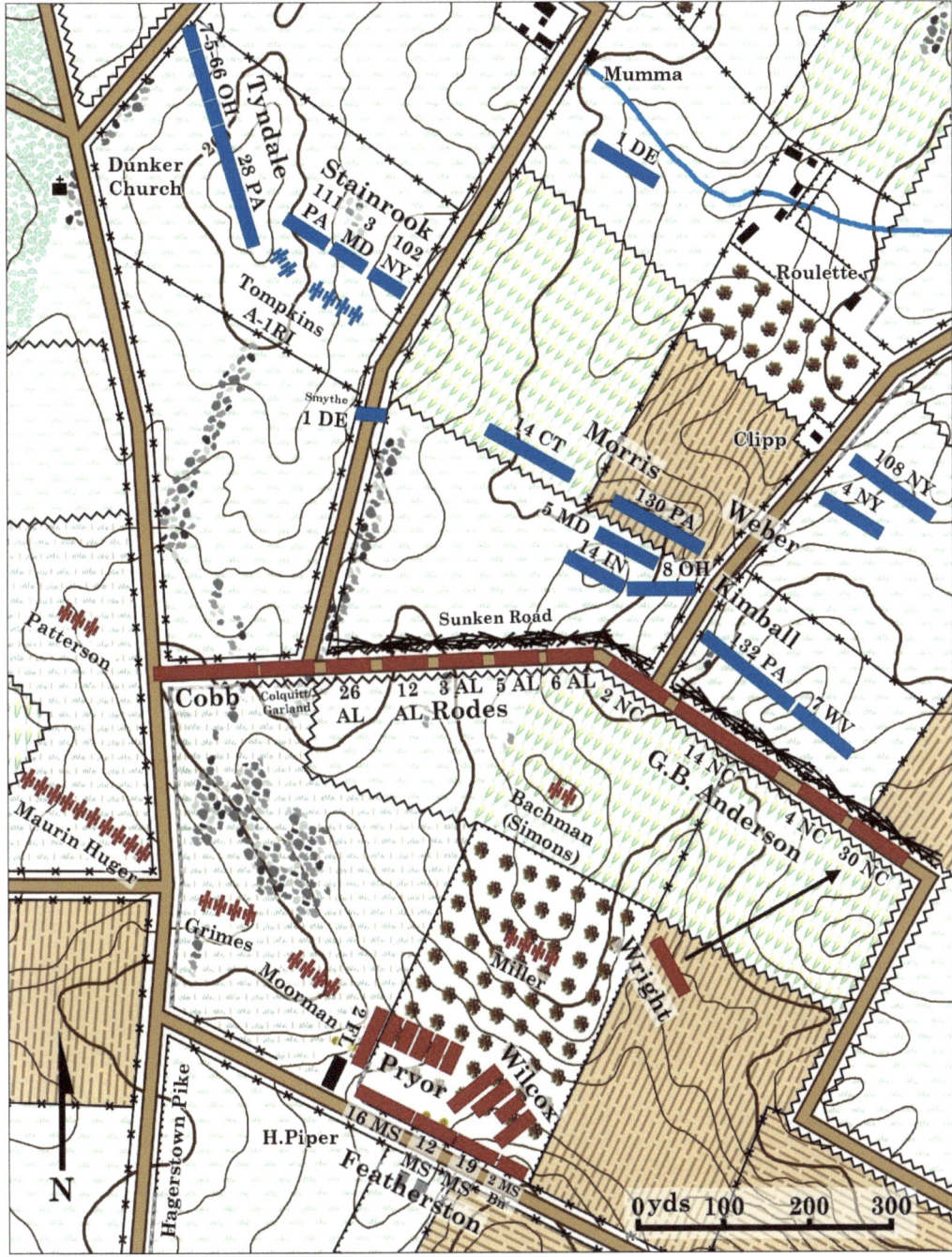

1. 10:20 a.m.
2. The inter-mixed Union brigades fall back to the crest of the hill. Both sides find cover and settle into a close range firefight.
3. Wright's Brigade moves forward to support G. B. Anderson. They suffer severe casualties crossing the exposed high ground in Piper's cornfield before reaching the Sunken Road. Wright is wounded before reaching the road. He remains on the field, but tactical control eventually falls to Colonel William Gibson of the 48th Georgia.[8]
4. The rest of Anderson's Division gathers in the Piper orchard. Anderson is wounded leading Wright into position, and Pryor takes command of the division. His leadership is disjointed and uncoordinated.[9]

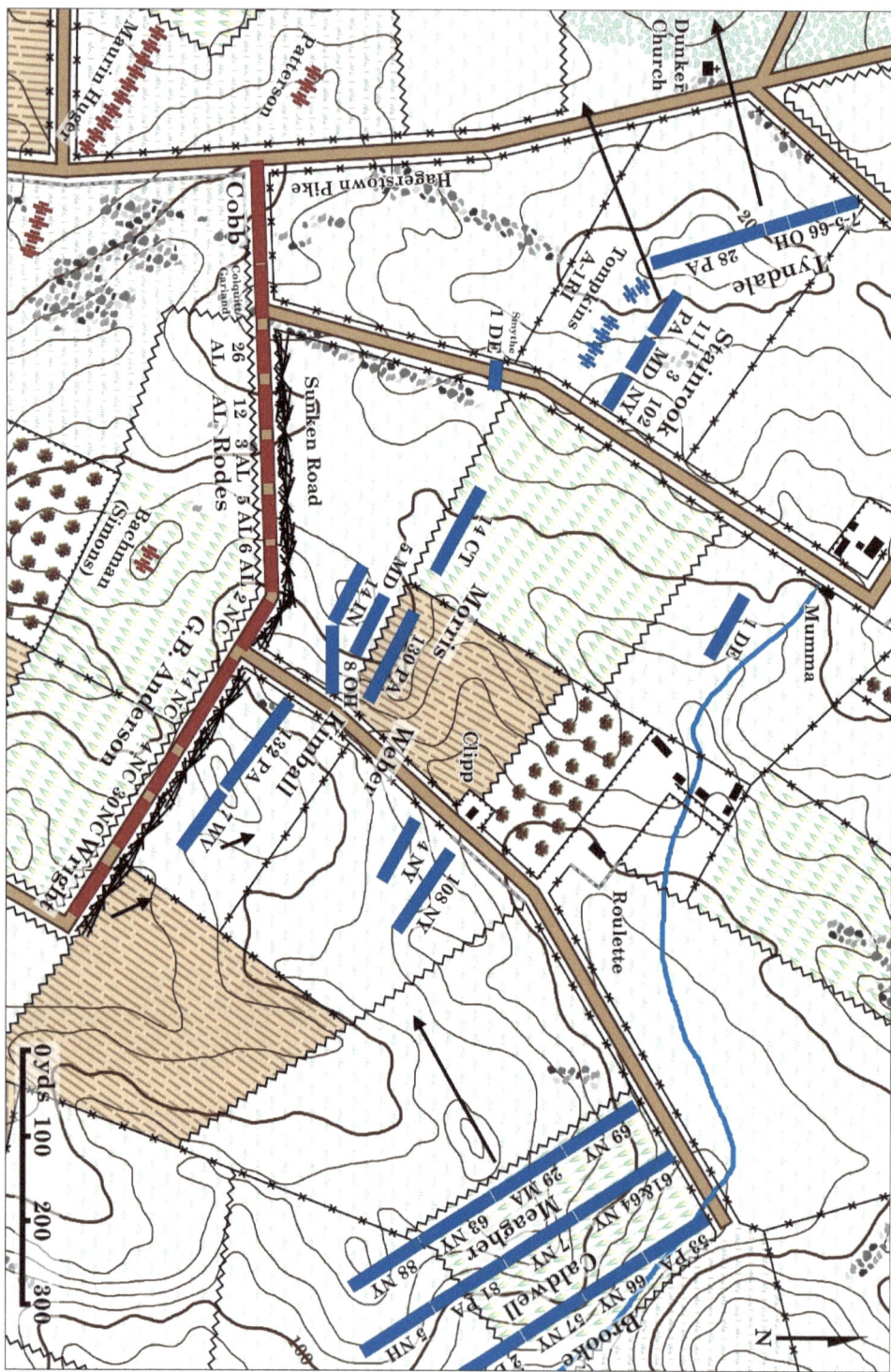

1. 10:30 a.m.
2. Wright, nominally exercising command though wounded, tries to order his brigade to charge and coordinate an advance with G. B. Anderson and Rodes. The only unit that is able to move forward is the 3rd Georgia from Wright's Brigade. They charge out of the Sunken Road in an attempt to flank the Union line. The 7th West Virginia changes front to meet them.[10]
3. Richardson's division arrives on the battlefield and moves toward the Sunken Road to support French.
4. Green's division advances into the West Woods.

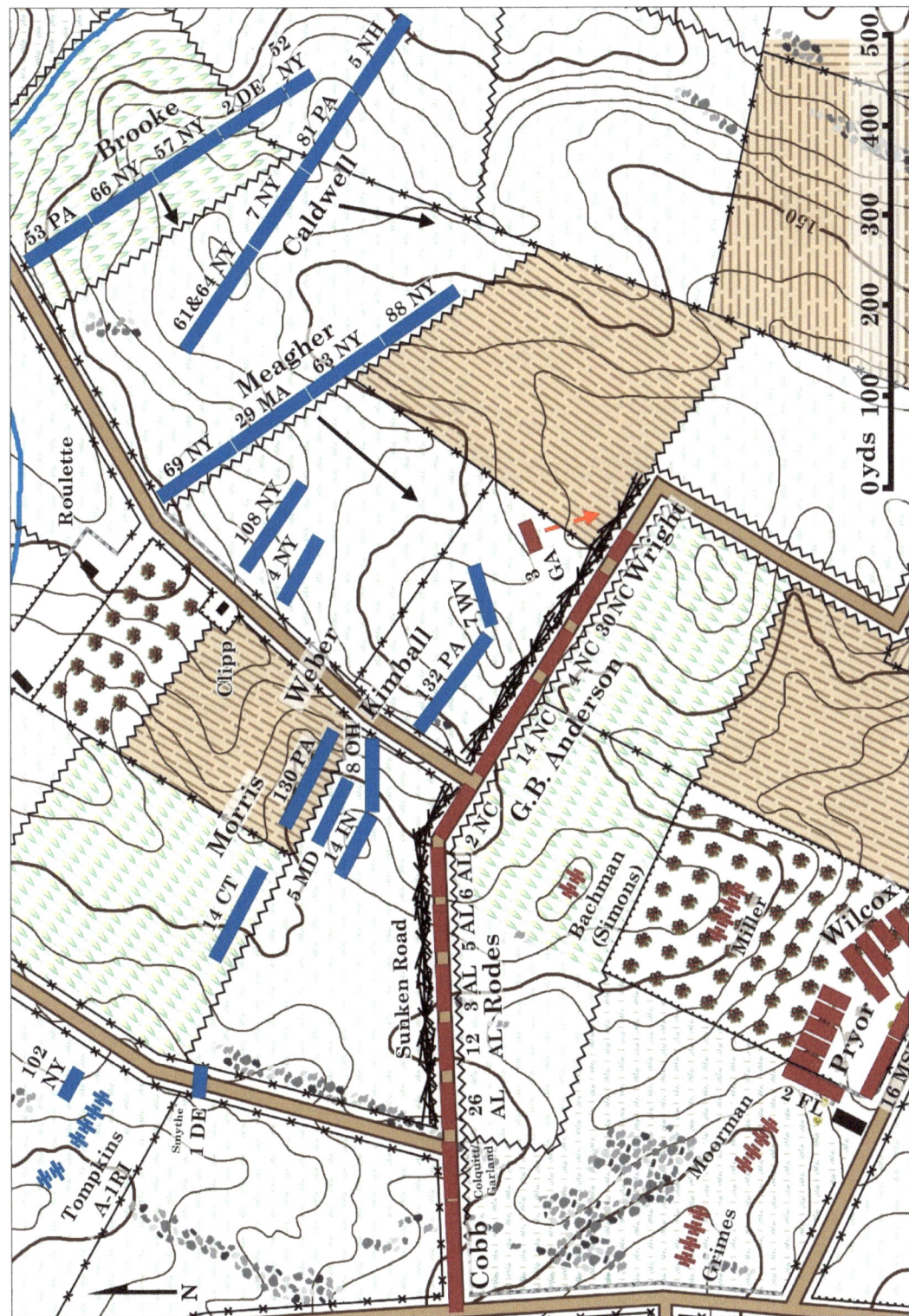

1. 10:40 a.m.
2. The 3rd Georgia retreats back to the Sunken Road in the face of resistance from the 7th West Virginia and Meagher's approaching Irish Brigade.[11]
3. The Irish Brigade moves south to help Kimball.
4. Caldwell's brigade shifts to the east instead of moving directly behind Meagher.

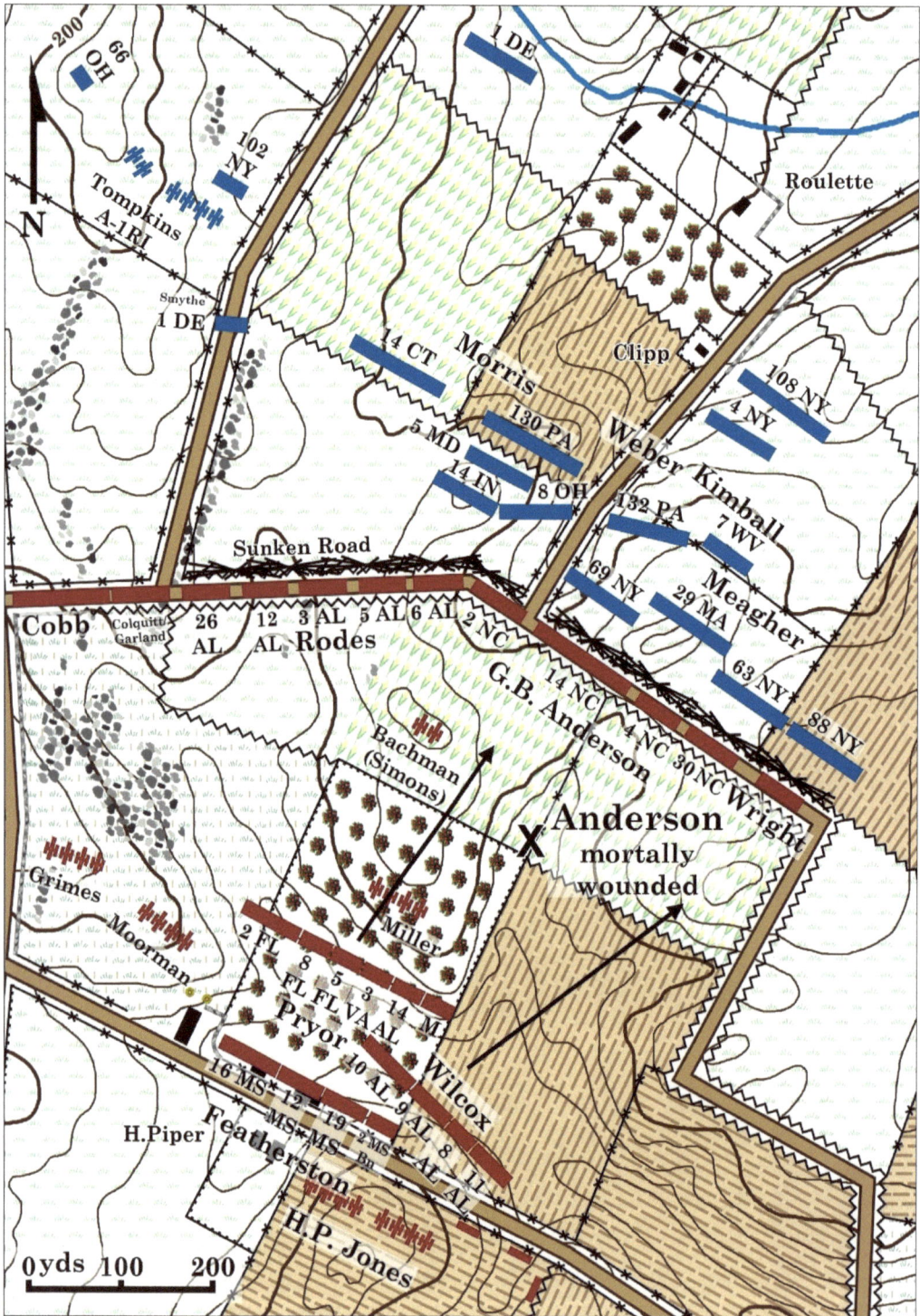

1. 10:45 a.m.
2. The Irish Brigade marches through the mix of French's units and assaults the Sunken Road. Their smoothbore muskets with buck and ball ammunition increase Confederate casualties.
3. Pryor's remaining brigades run the gauntlet through the Piper cornfield and reinforce the Sunken Road.
4. Brigadier General George B. Anderson is mortally wounded returning to his brigade after conferring with his division commander, Daniel H. Hill. Colonel Risden T. Bennett of the 14th North Carolina takes command.[12]

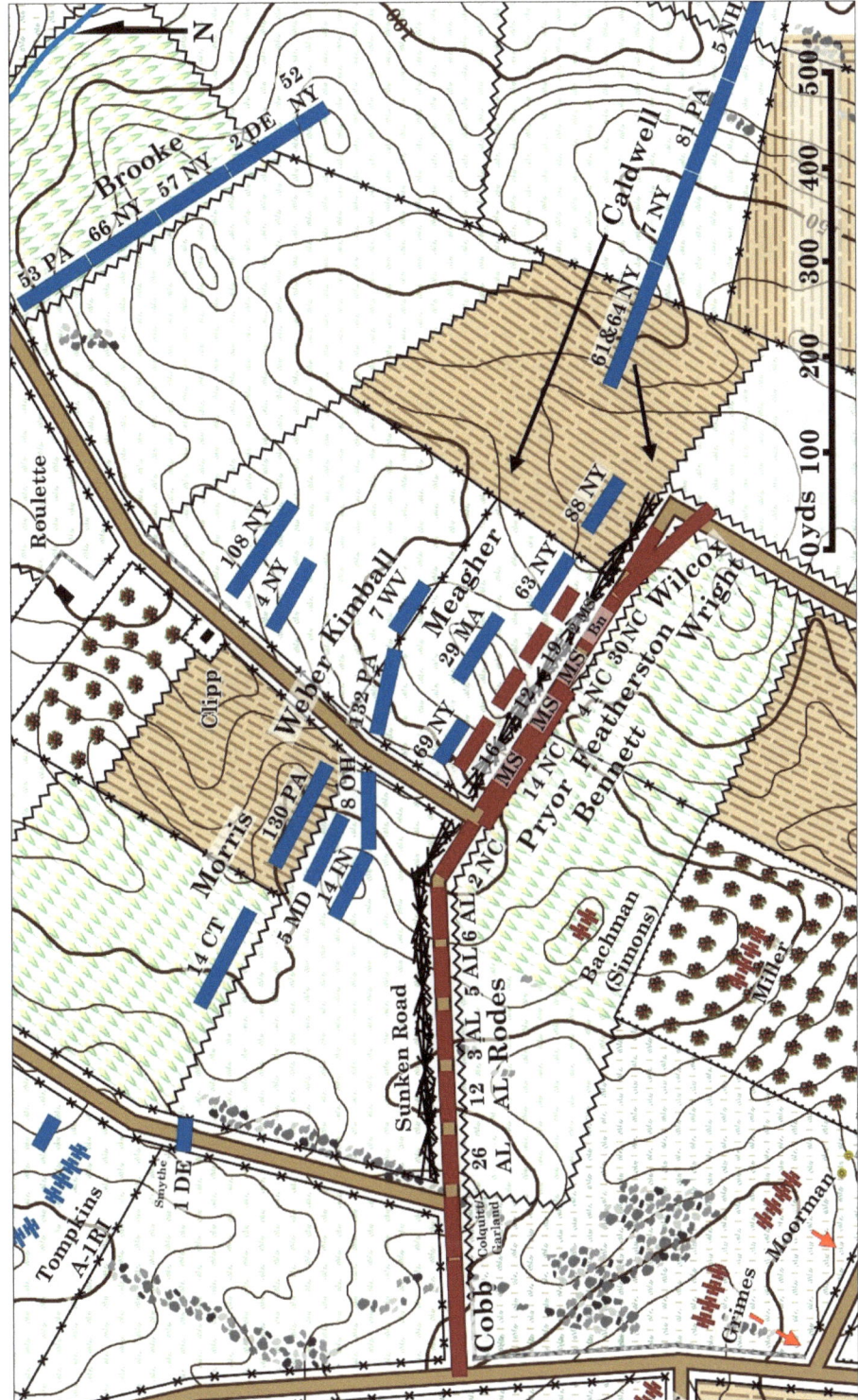

1. 10:50 a.m.
2. Featherston's Brigade charges out of the Sunken Road, engages the Irish Brigade, but falls back after about five to ten minutes.[13]
3. Caldwell's brigade approaches from the east. The consolidated 61st & 64th New York forms on the left of the Irish Brigade. Instead of flanking the Confederate line, the rest of Caldwell's regiments continue west and form under the hill behind Meagher.[14]

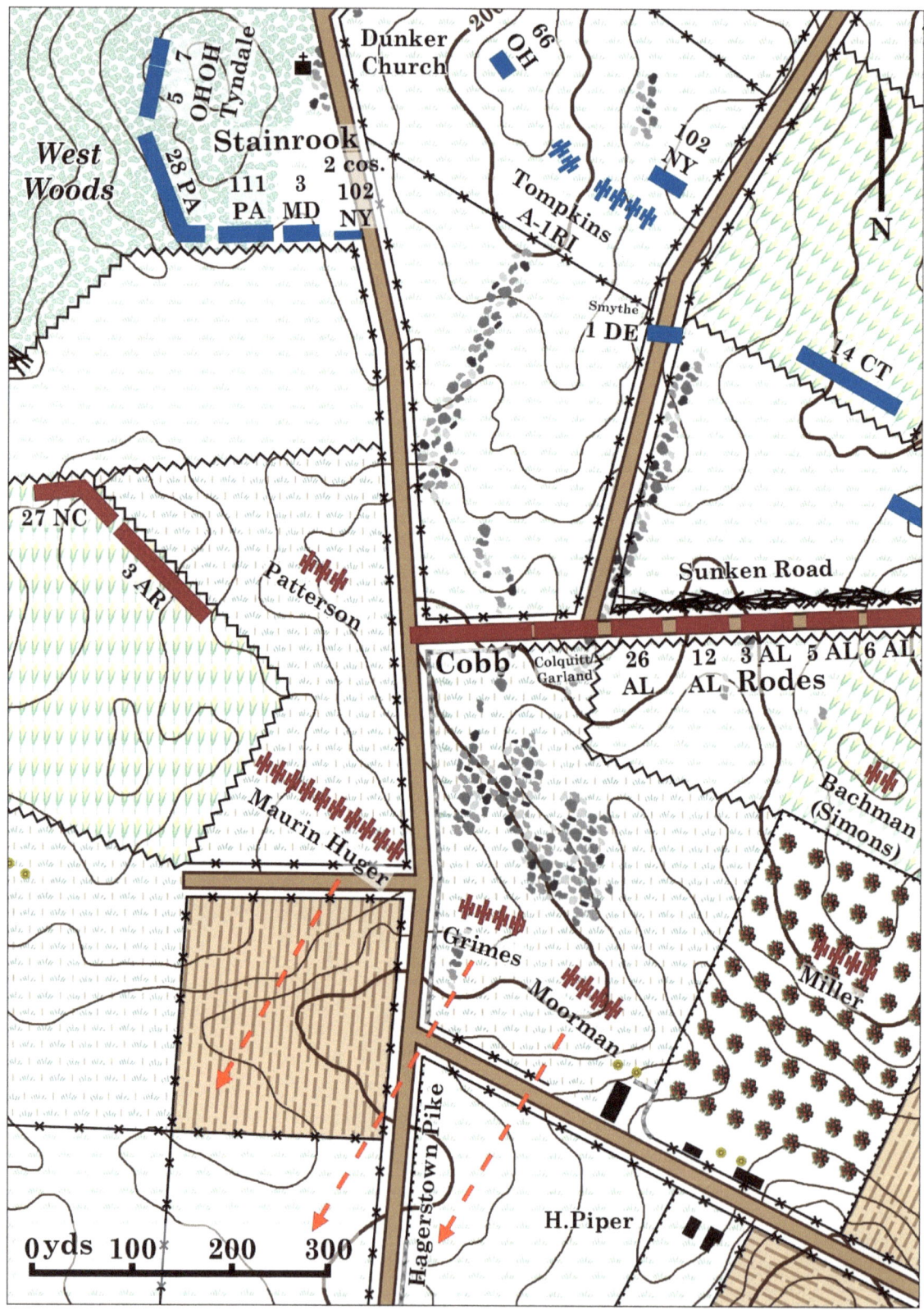

1. 10:50 a.m.
2. Saunders' Battalion, the artillery for Anderson's Division, is forced to retire. They have been effectively silenced by accurate fire from Battery A, 1st Rhode Island.

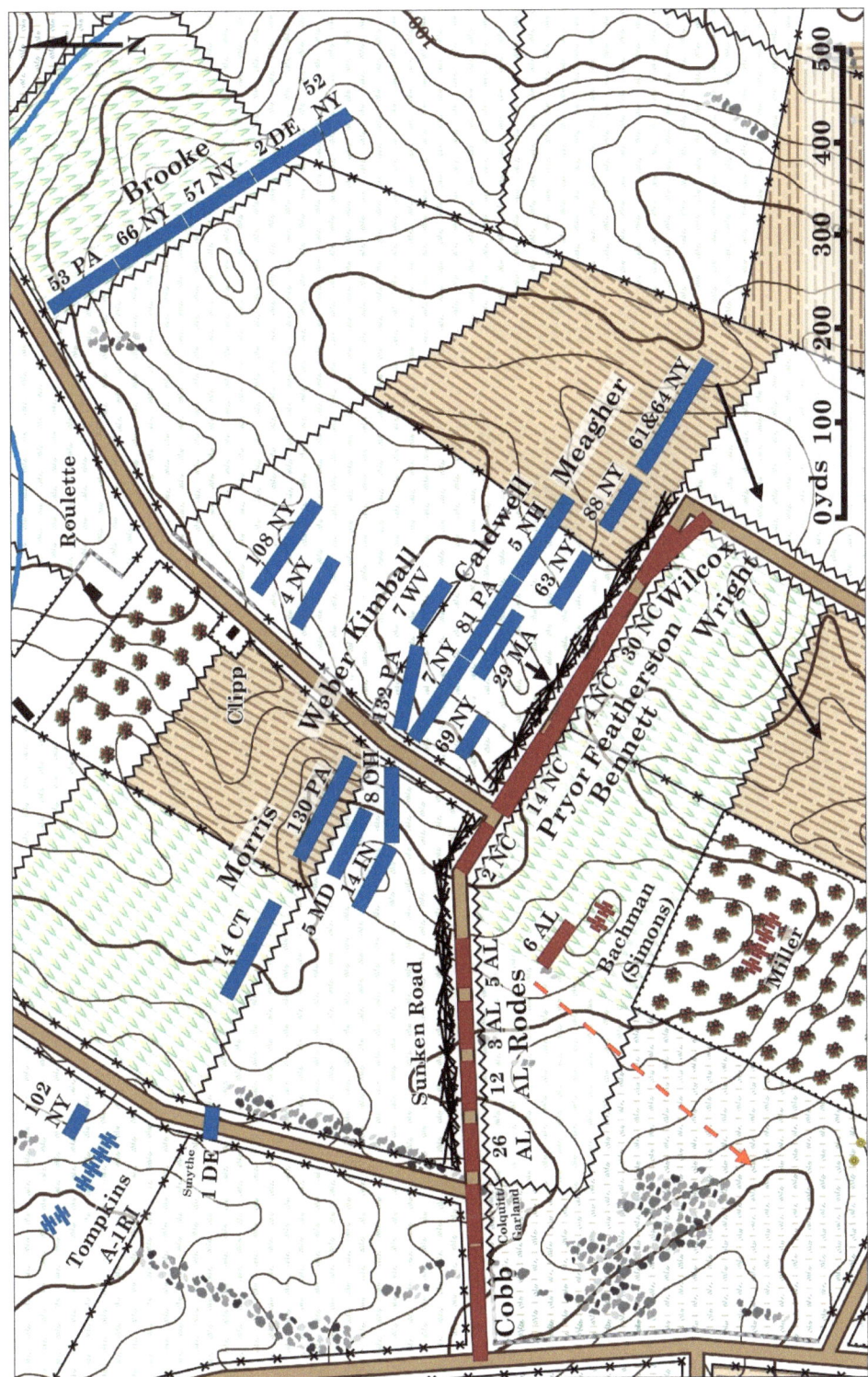

1. 11:00 a.m.
2. The commander of the 6th Alabama, misunderstanding an order from Rodes, falls back from the Sunken Road. The rest of the brigade think it is an order to retreat, and the brigade leaves the road and moves south despite Rodes' best efforts.[15]
3. The 61st & 64th New York advance to the high ground at the bend in the Sunken Road.[16]
4. Wright pulls his brigade back.

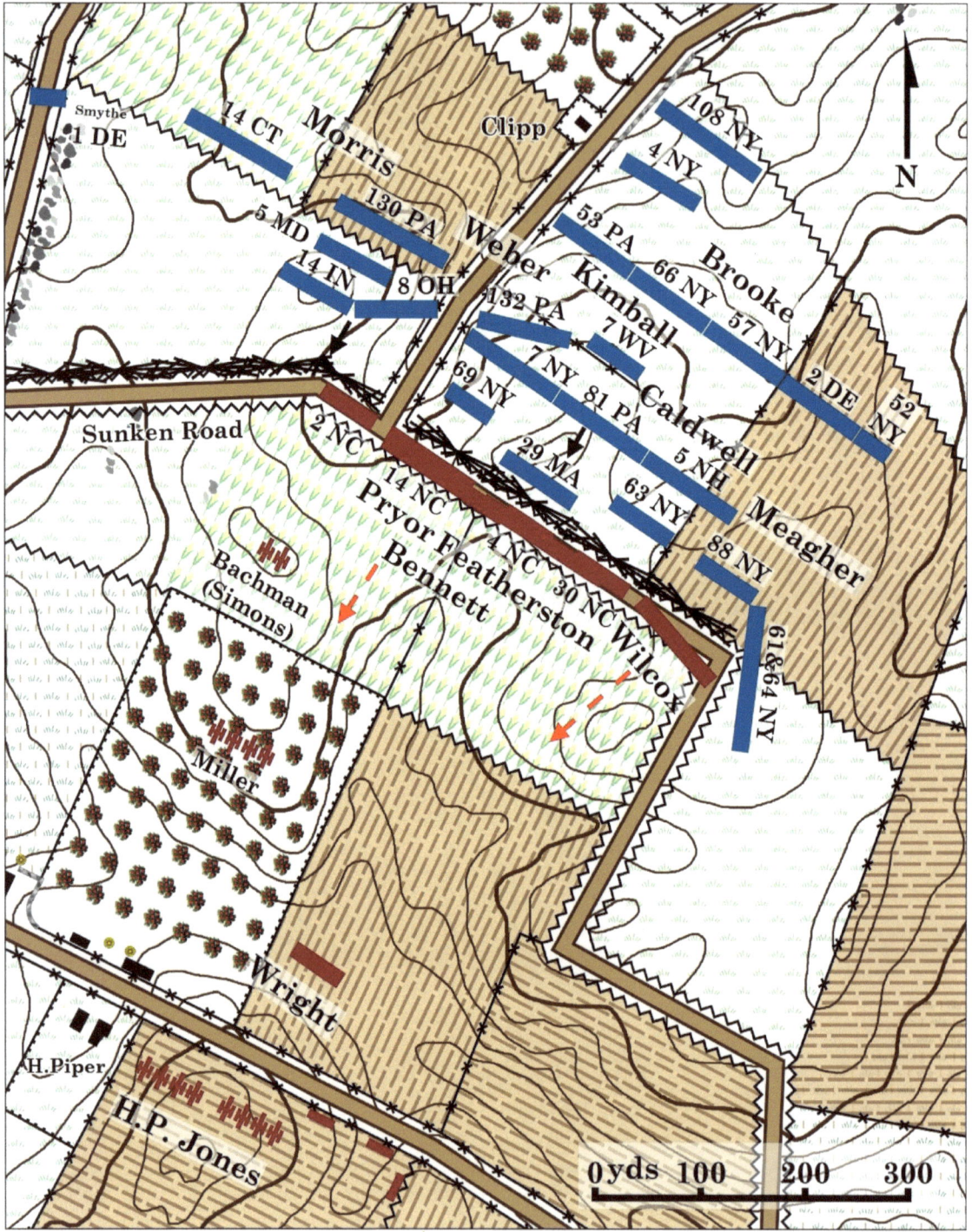

1. 11:10 a.m.
2. Seeing Rodes leave, Kimball orders the right of his brigade forward into the gap.[17]
3. The 61st & 64th New York fire down the length of the Sunken Road. Many Confederates begin to surrender or retreat.[18]
4. Caldwell orders his brigade to charge through the thin lines of the Irish Brigade and into the Sunken Road.[19]
5. Pryor's division (R. H. Anderson) retreat from the Sunken Road, carrying the right of George B. Anderson's Brigade with it.

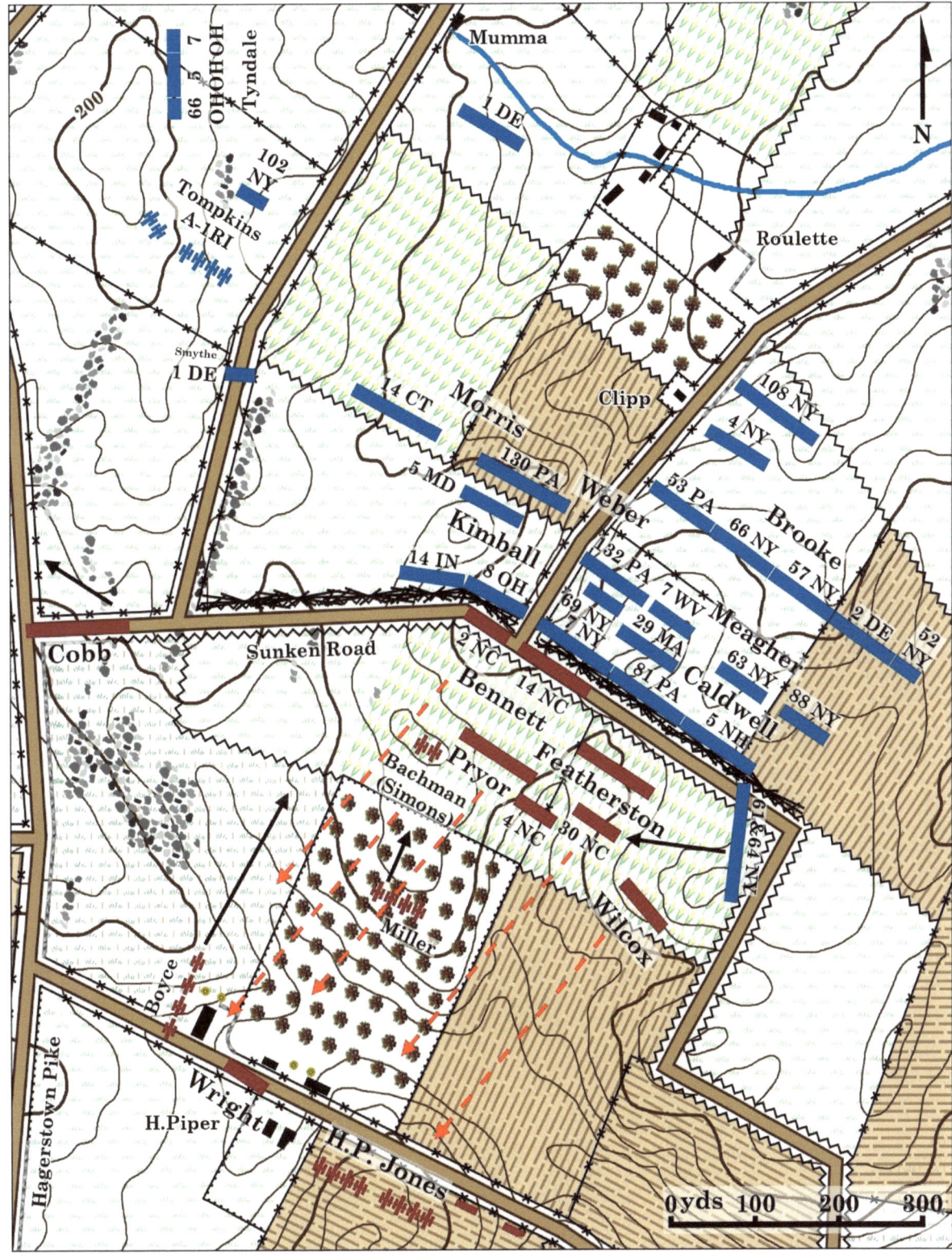

1. 11:15 a.m.
2. The remaining defenders of the Sunken Road rout and everyone falls back to the Piper farm.
3. The 61st & 64th New York advances down the roadway, largely unopposed.[20]
4. Cobb's Brigade changes front to face east.[21]

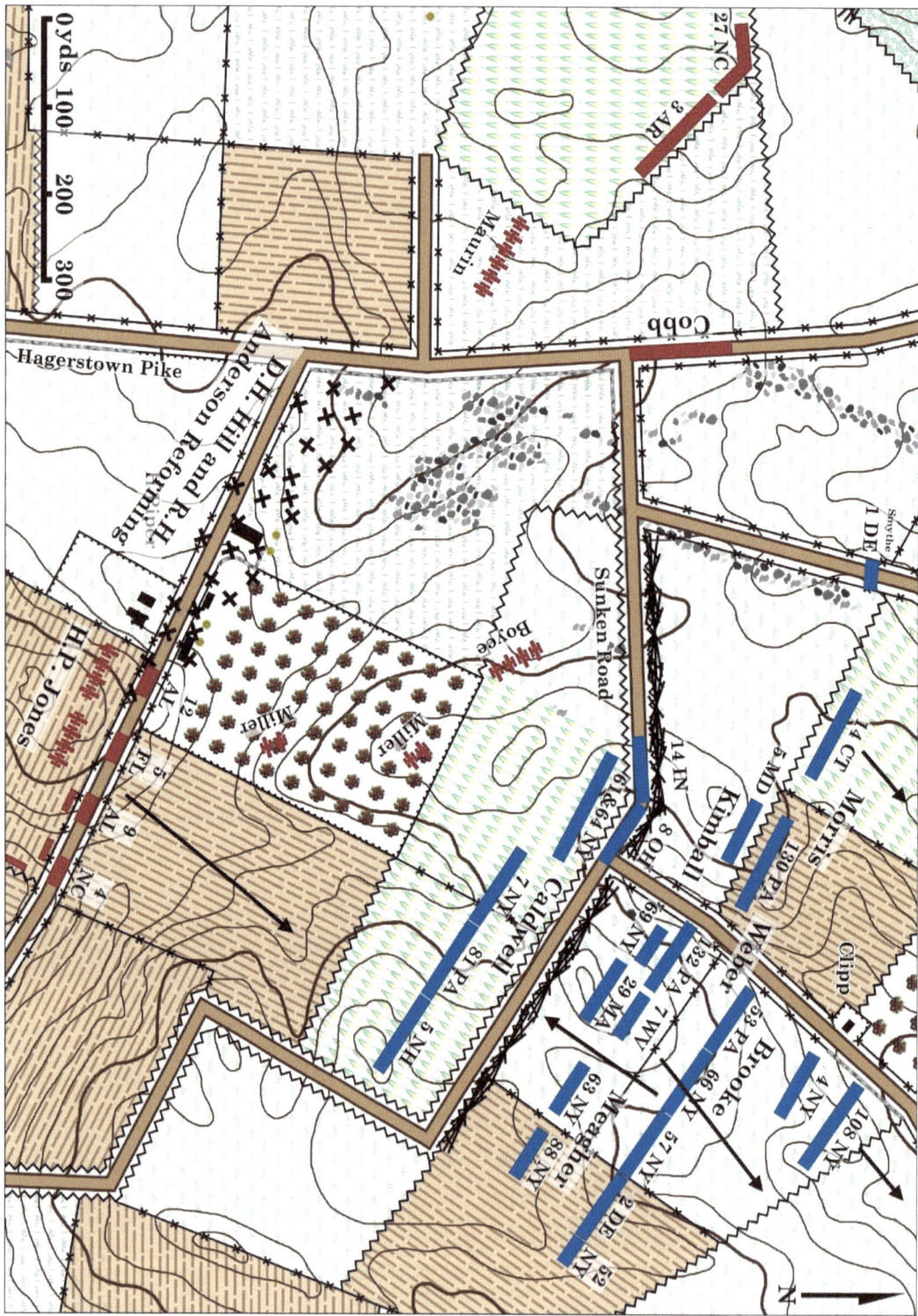

1. 11:30 a.m.
2. Brook's brigade moves forward.
3. The Irish Brigade falls back from the Sunken Road.
4. The divisions of D. H. Hill and Pryor (R. H. Anderson) attempt to rally near the Piper farm lane.
5. A collection of Confederate regiments and fragments from different commands moves forward to counterattack Caldwell in the Piper cornfield.[22]

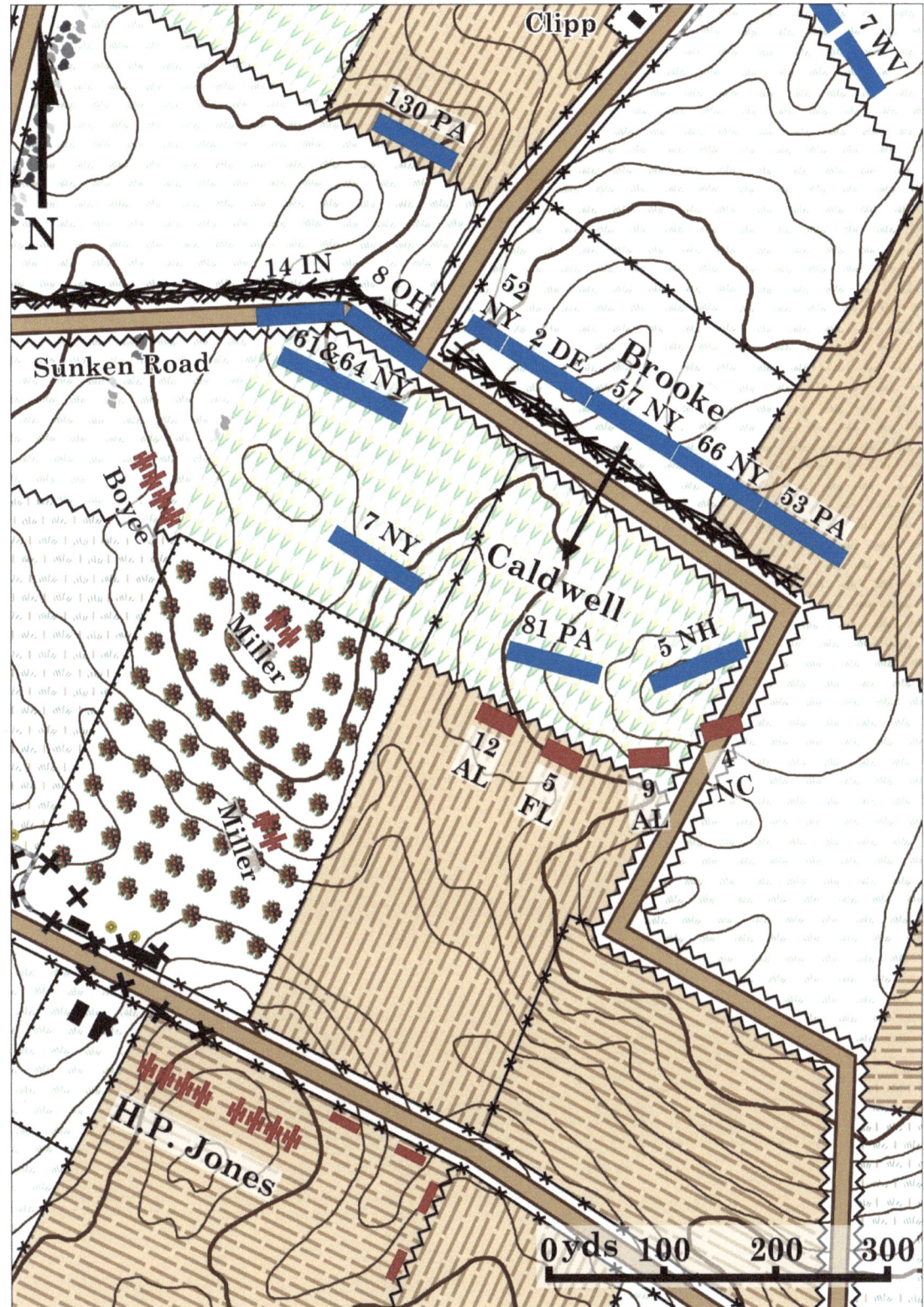

1. 11:45 a.m.
2. Caldwell fights the Confederate counterattack in the Piper cornfield.[23]
3. Brooke advances his brigade south to assist.
4. D. H. Hill and Pryor try to rally their commands near the Piper farm.

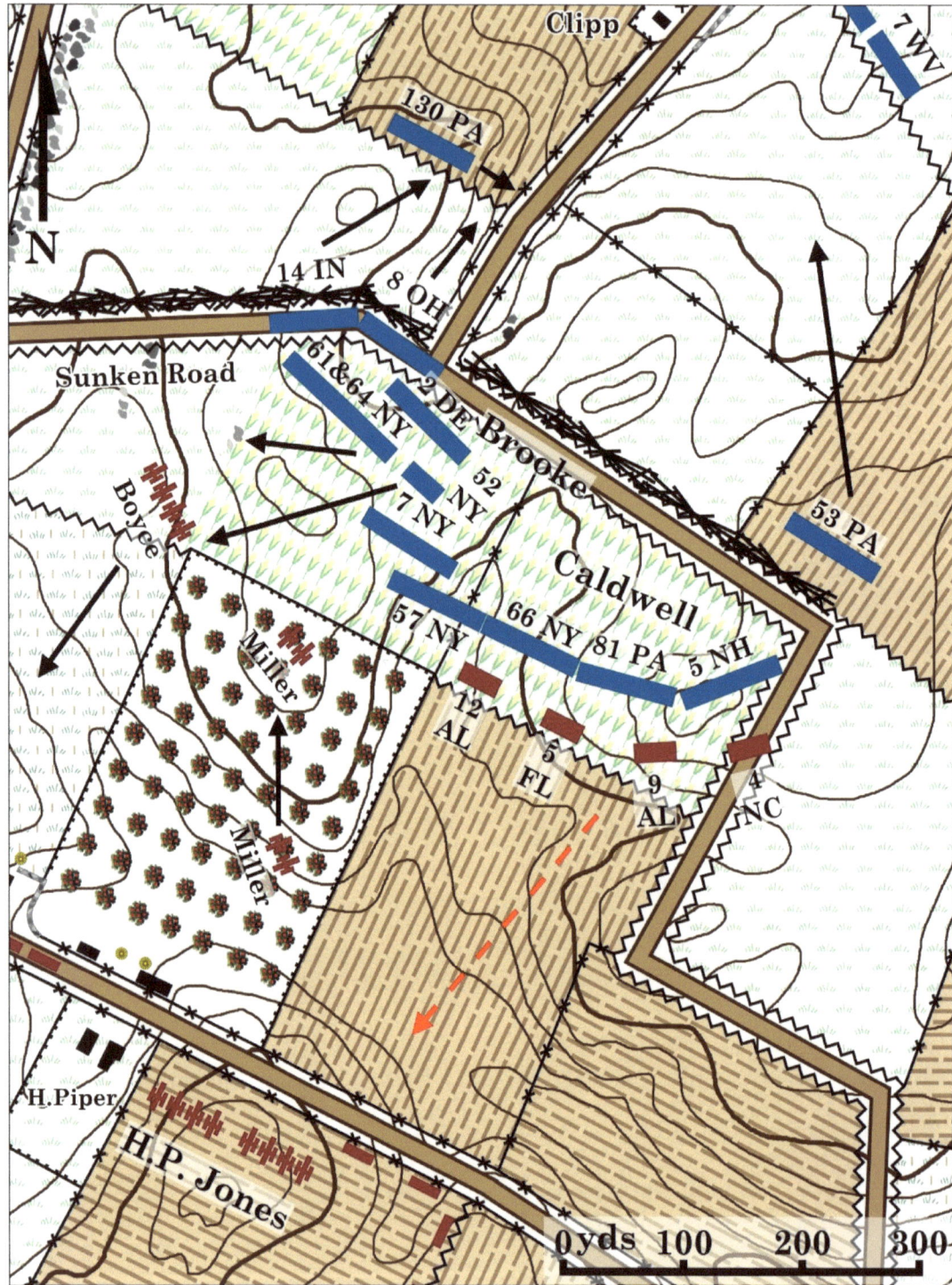

1. 12:00 p.m.
2. Brooke arrives to assist Caldwell, forcing the Confederates in the Piper cornfield to retreat.
3. Kimball pulls back from the Sunken Road in response to Confederate threats to the west. This is Cobb and Cooke's two regiments clearing the plateau above the Dunker Church.[24]
4. Brook likewise sends regiments to face west along the edge of the cornfield. The 61st and 64th New York does the same. The 53rd Pennsylvania heads north.[25]
5. Boyce's Battery is forced back by these movements.

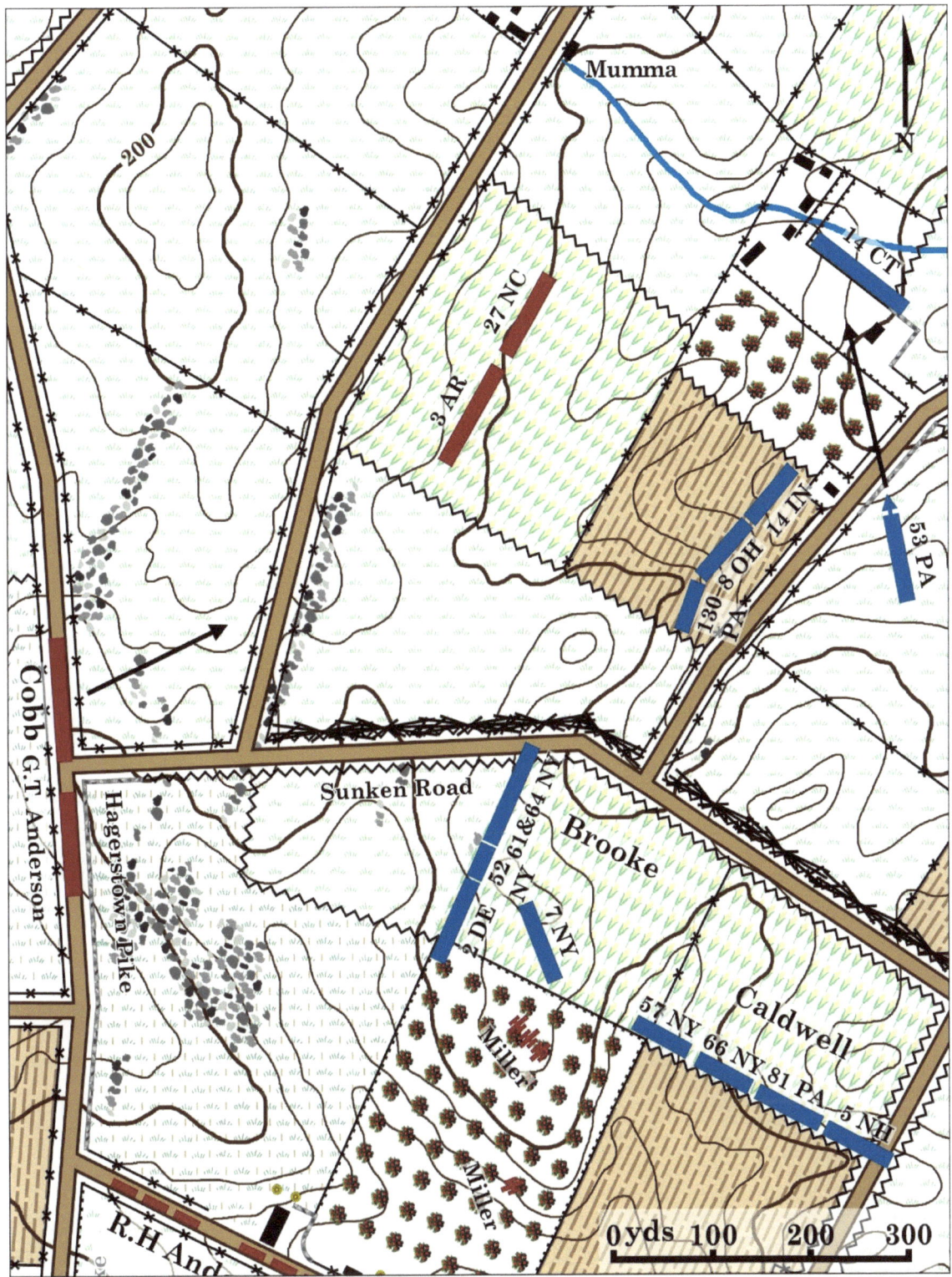

1. 12:15 p.m.
2. Cooke charges over the plateau and into the Mumma cornfield. They engage Kimball stationed west of the Clipp house.[26]
3. The 53rd Pennsylvania arrives at the Roulette farm and moves west to confront Cooke.[27]
4. D. H. Hill orders Cobb's Brigade to advance.[28]

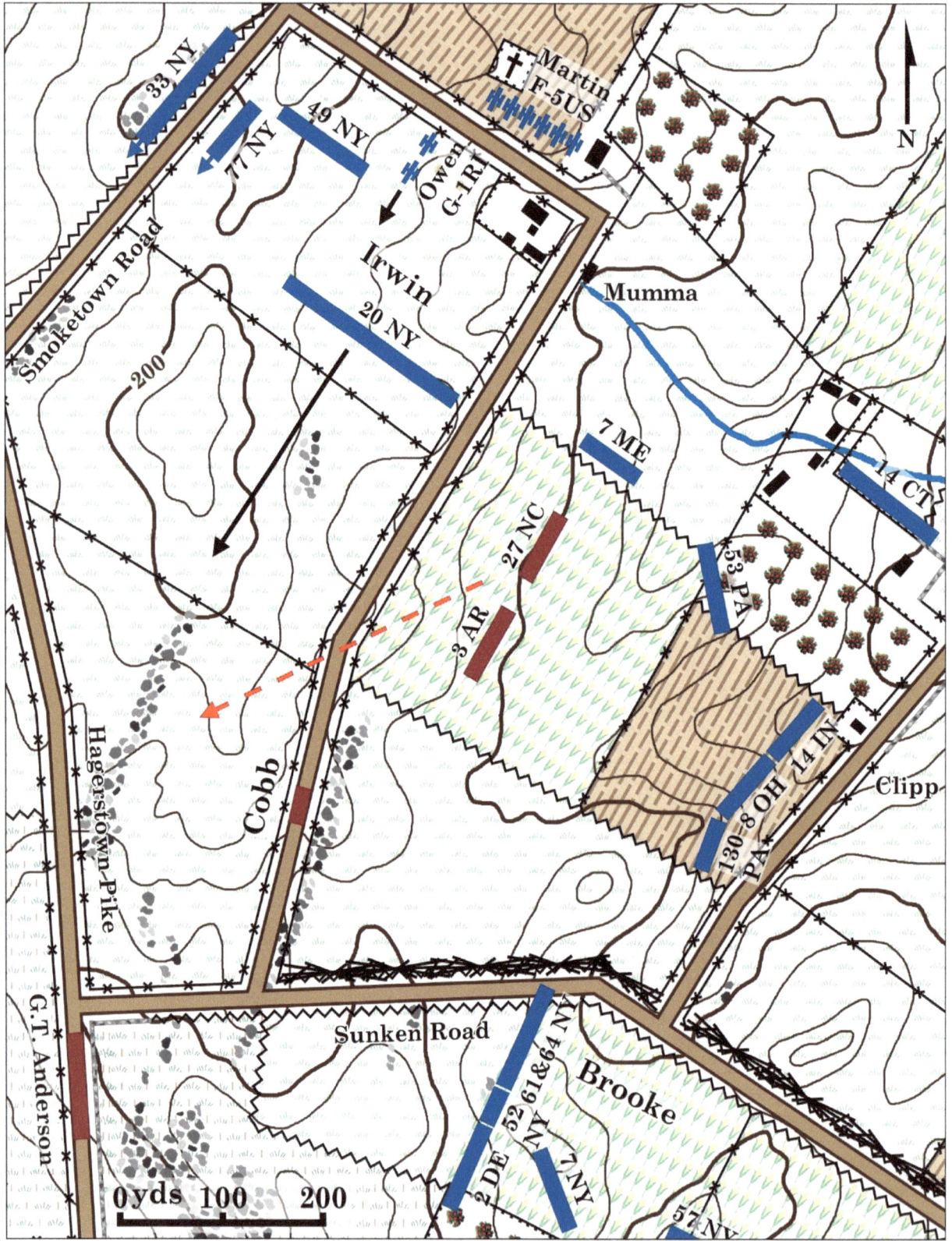

1. 12:20 p.m.
2. Irwin's brigade from the fresh Sixth Corps advances south from the Mumma farm into Cooke's left flank. Facing Kimball to the east and Irwin to the north, Cooke retreats.[29]

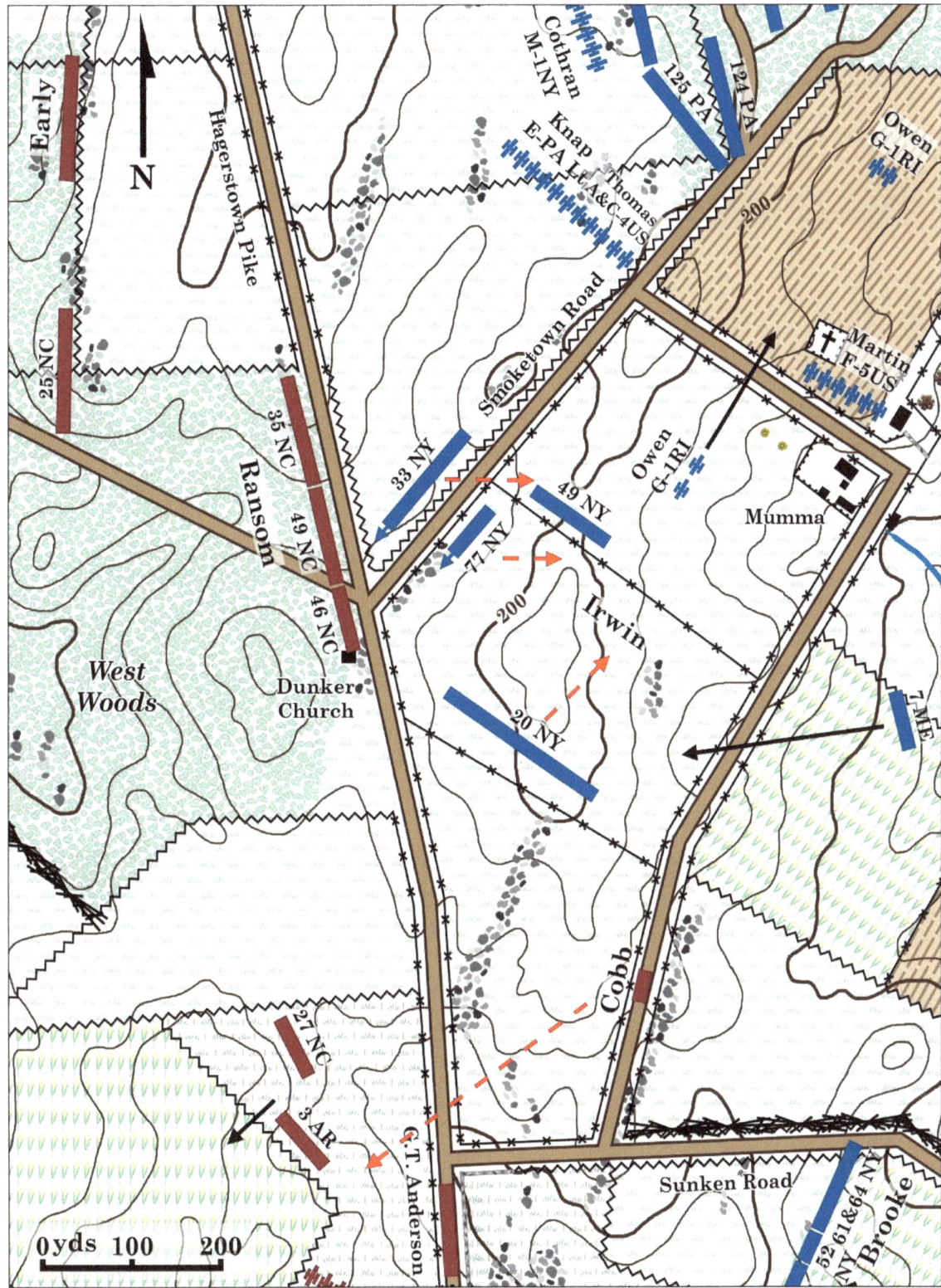

1. 12:30 p.m.
2. Irwin continues advancing toward the West Woods. Some regiments remain in column.
3. Ransom's Brigade fires several volleys from the wood's edge into the surprised Union regiments. They hastily fall back to the other side of the plateau.[30]

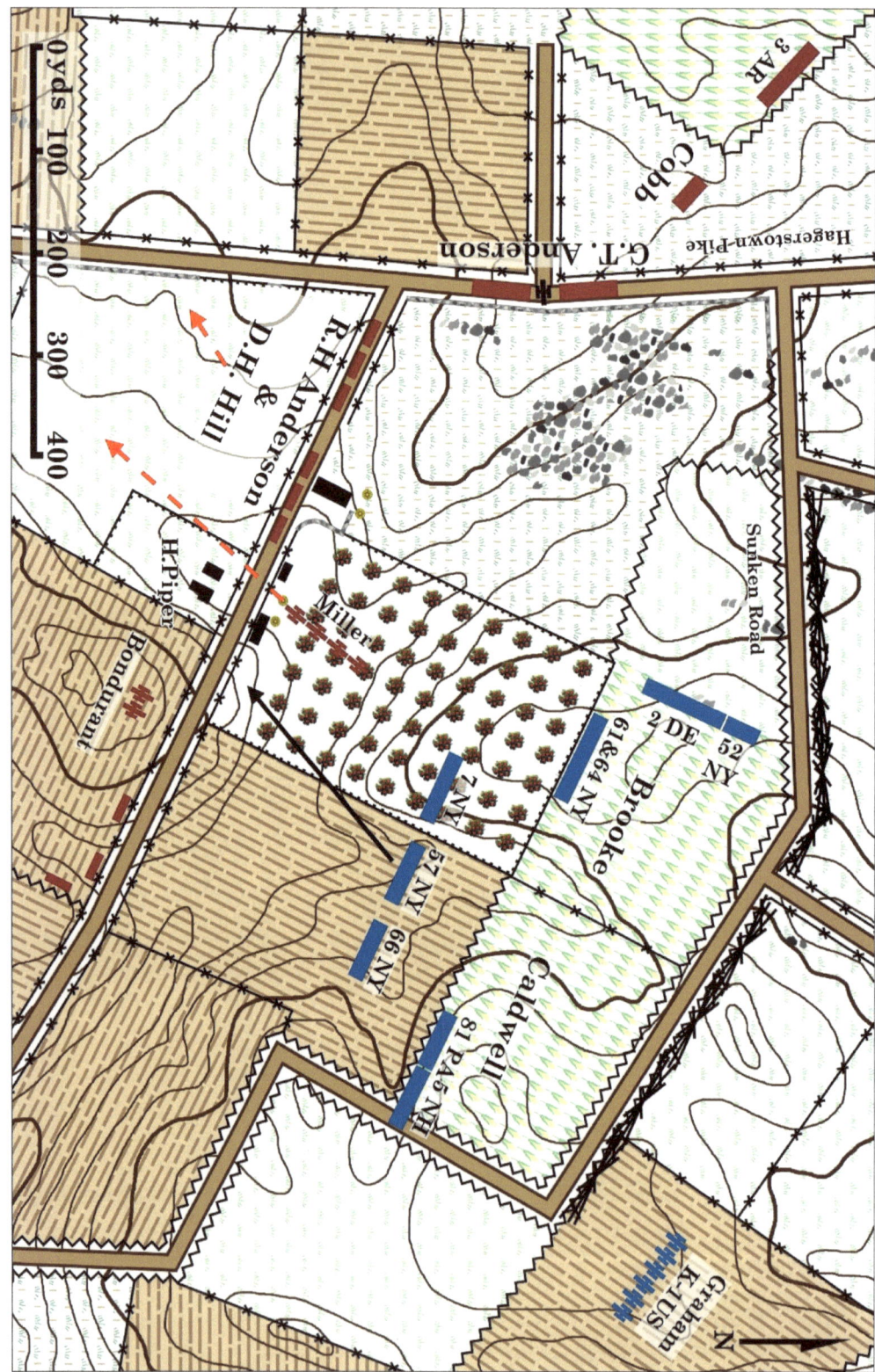

1. 12:30 p.m.
2. The 7th New York from Brooke's brigade and the 57th and 66th New York from Caldwell advance south toward the Piper farm.[31]
3. Miller's Battery is forced to withdraw.

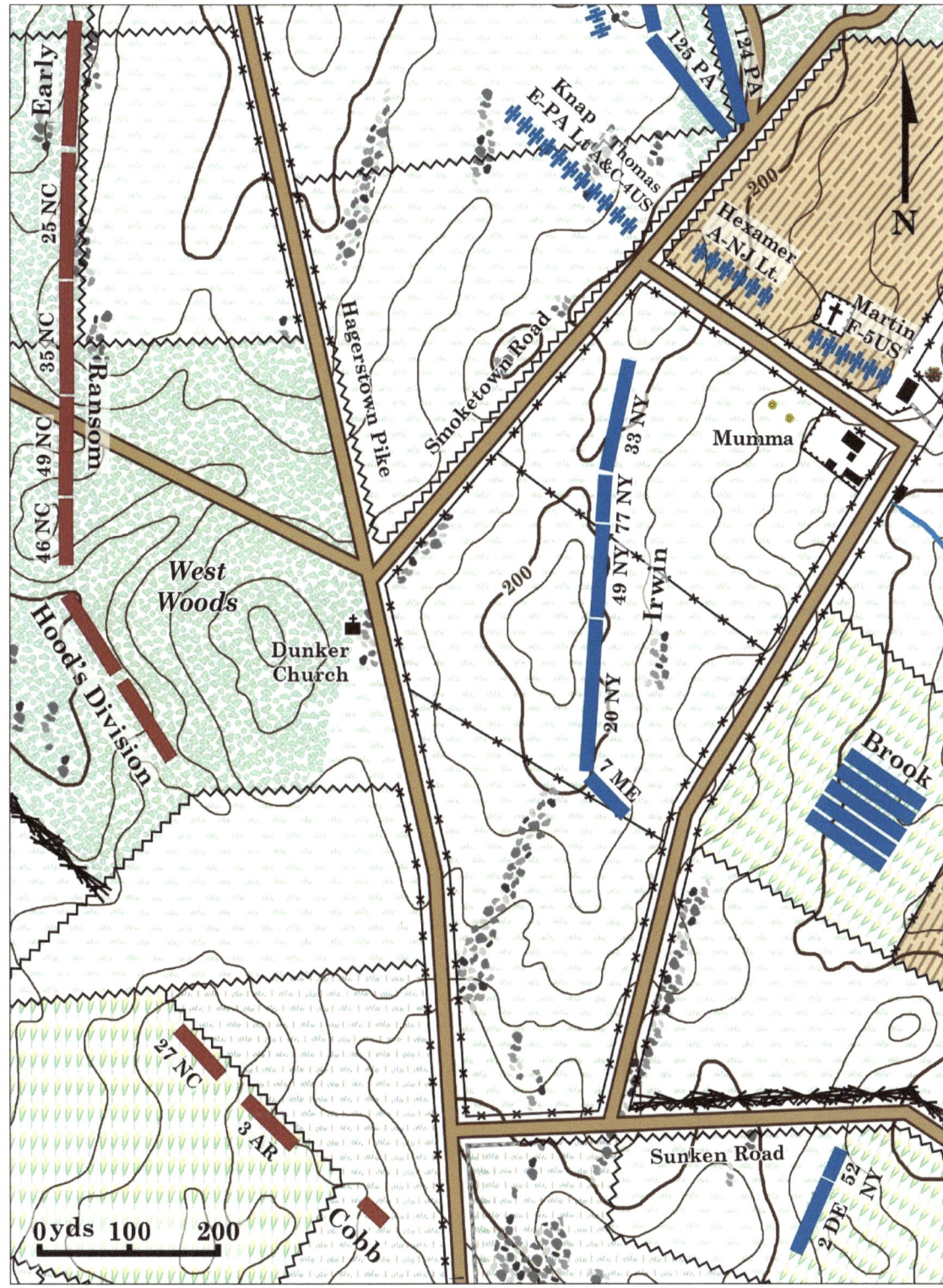

1. 12:45 p.m.
2. Irwin's brigade reforms on the reverse slope of the plateau.
3. Ransom falls back from the edge of the West Woods.
4. Brook's brigade moves forward and deploys on Irwin's left at the southern edge of the Mumma cornfield.

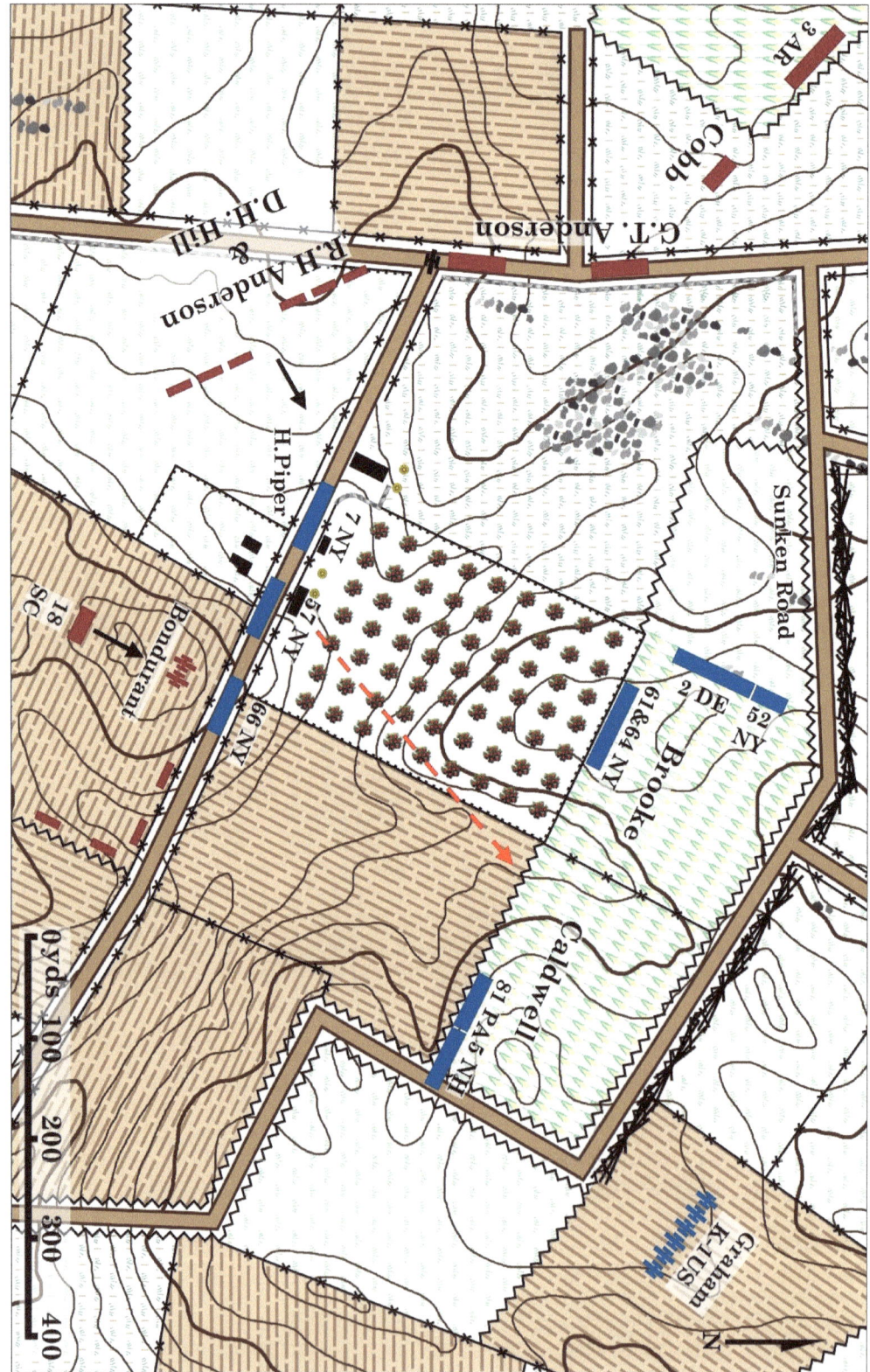

1. 12:45 p.m.
2. The advancing regiments reach the Piper farm lane.
3. After a short stay, the regiments fall back. They are isolated and exposed, and receiving fire from three sides.[32]

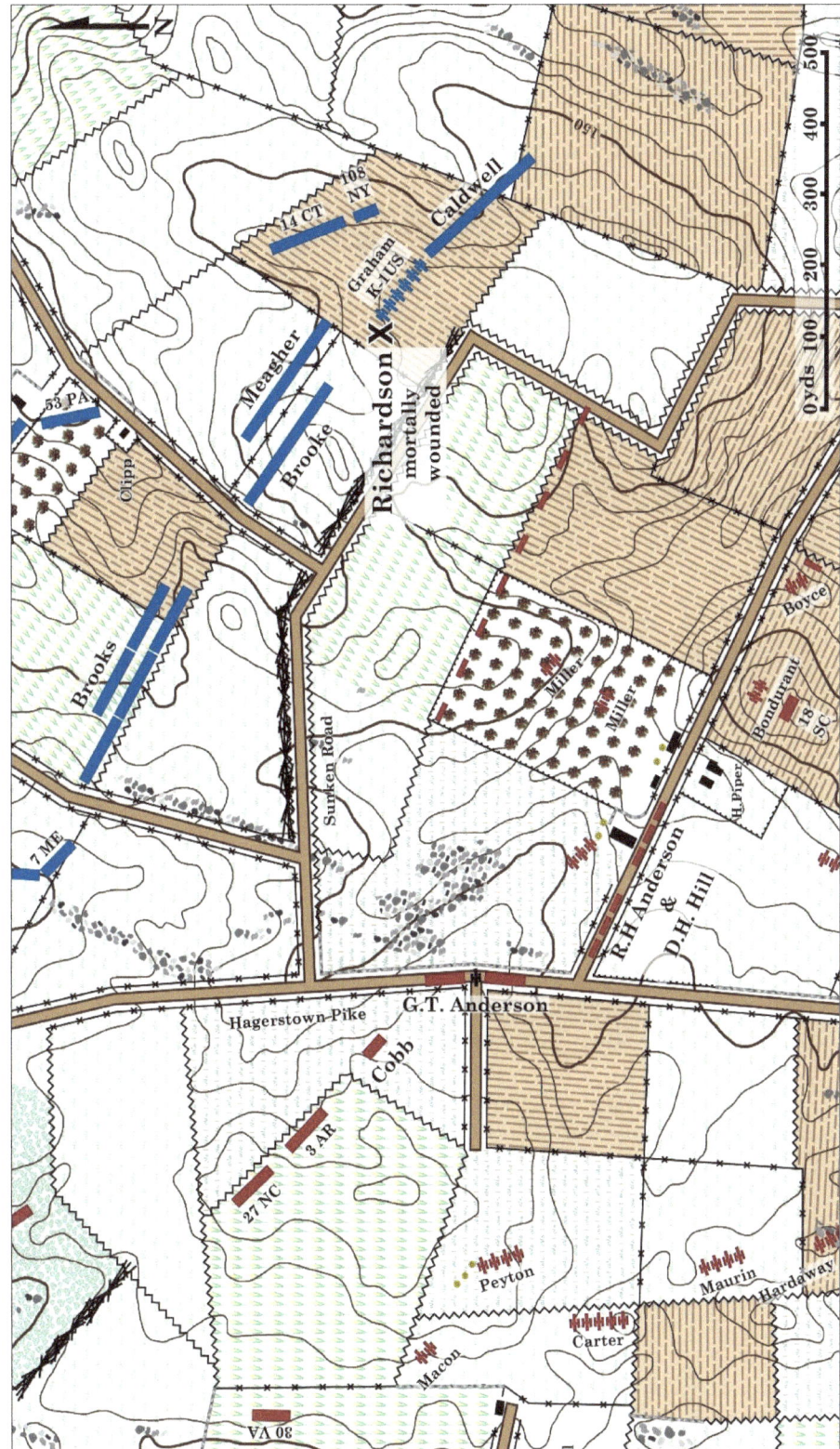

1. 1:00 p.m.
2. The lines around the Sunken Road settle into a stalemate.
3. Major General Israel B. Richardson is mortally wounded while attending to Battery K, 1st United States.[33]

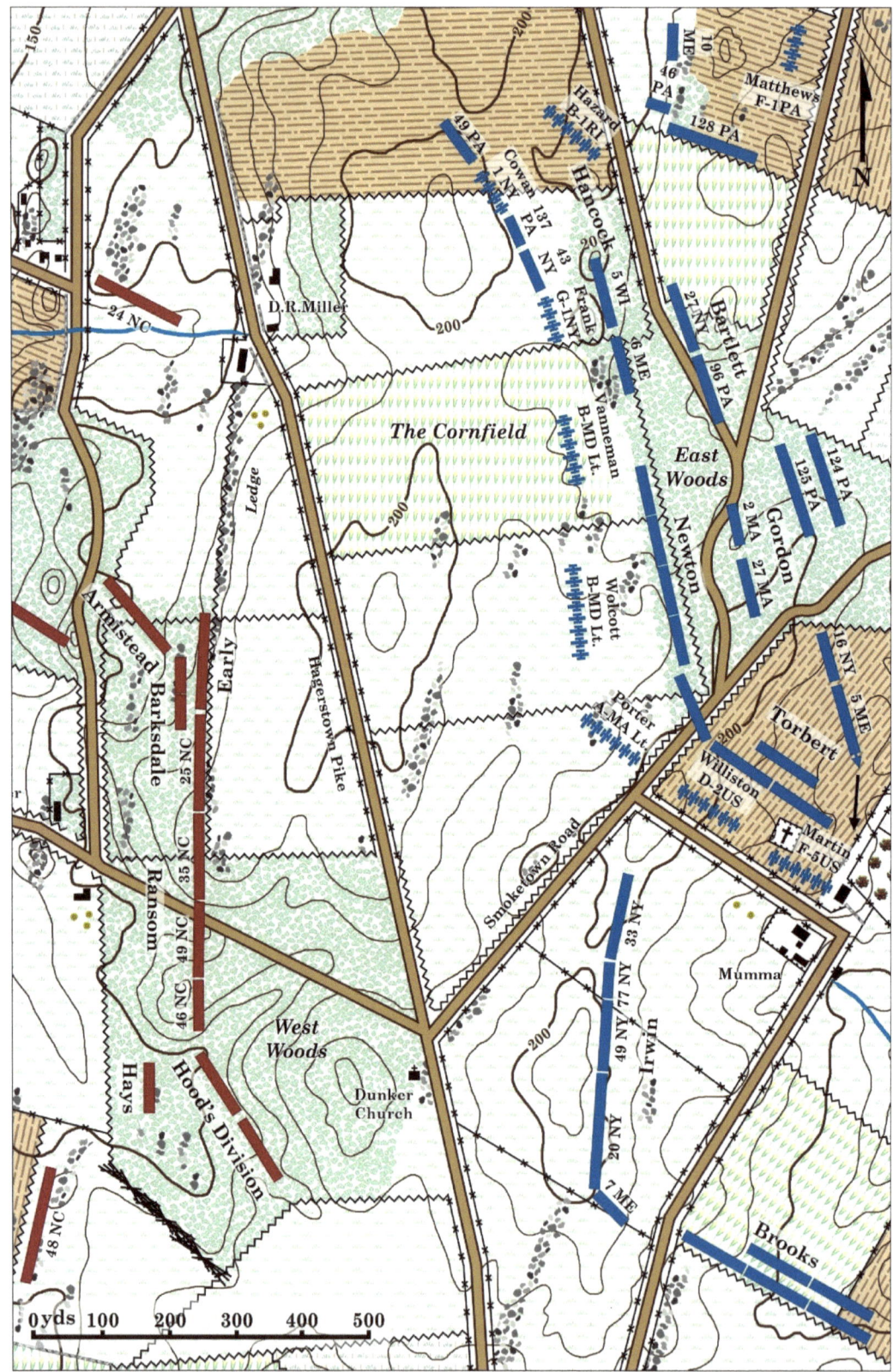

1. 3 p.m.
2. The northern front during the afternoon. Both sides have settled into an uneasy stalemate on either side of the high ground along the Hagerstown Turnpike.

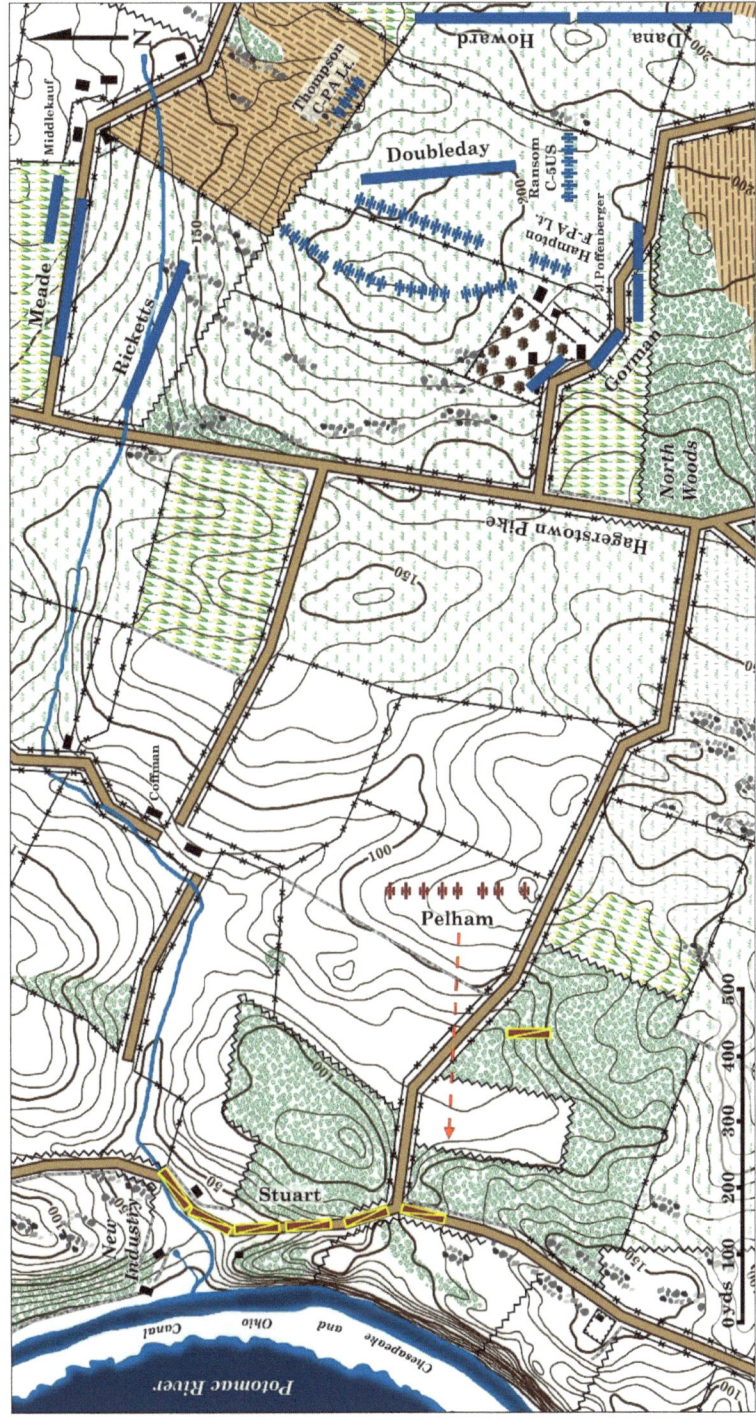

1. 3:30 p.m.
2. Jackson orders Stuart to send a force around the Union right and drive the enemy into the Potomac.[34]
3. Stuart marches to New Industry along the river. At approximately 3:30 he sends a mix of artillery under Pelham to the high ground east to bombard the Union.[35]
4. The massed Union batteries around the J. Poffenberger house silence and drive off the Confederate artillery in about 10 minutes.
5. Stuart abandons the effort and nonsensically tells Jackson the Union line extends to the Potomac and cannot be turned.[36]

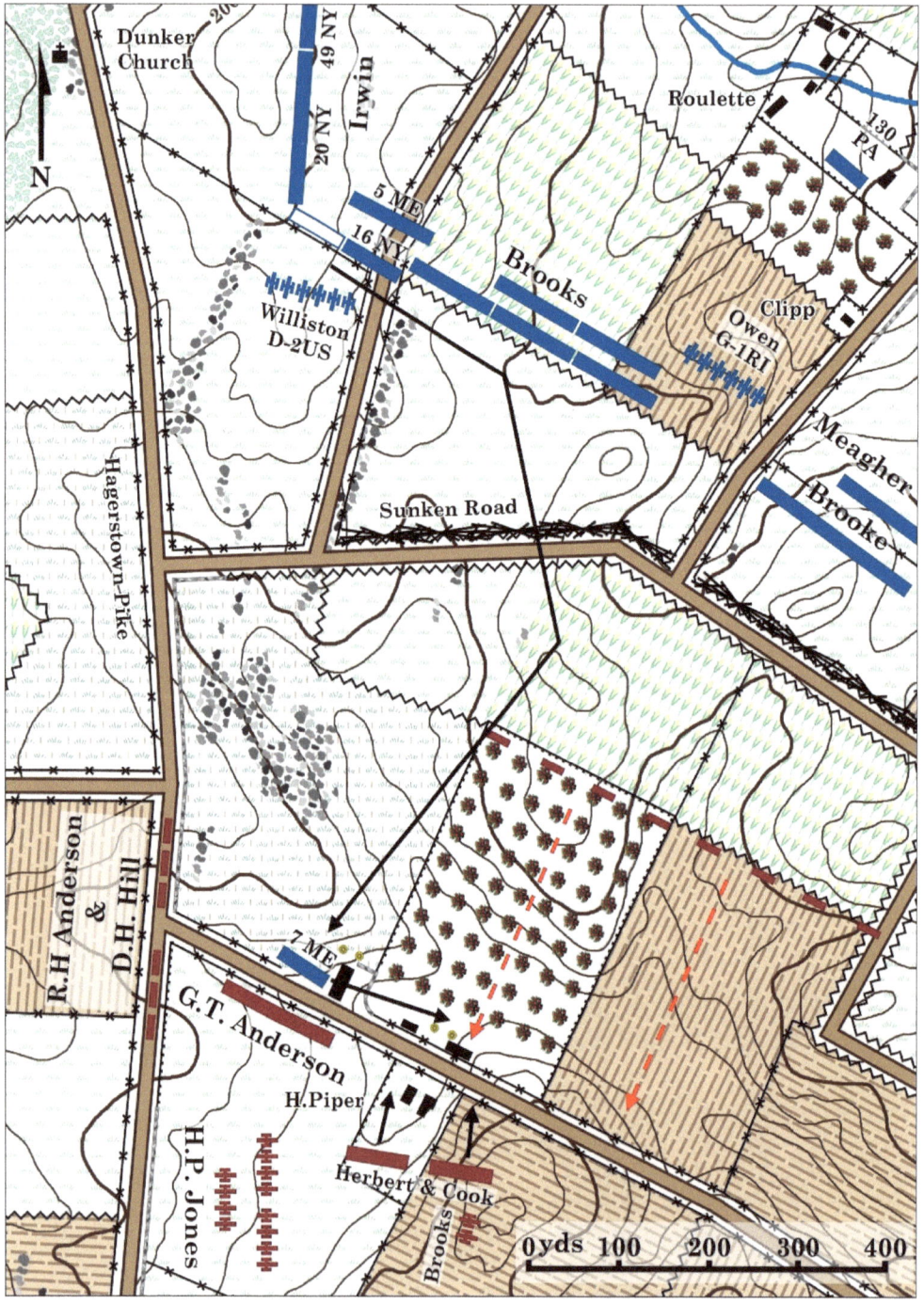

1. 5 p.m.
2. Irwin orders the 7th Maine to drive the enemy skirmishers from the Piper orchard and farm buildings.[37]
3. The regiment advances east, then turns south over the Sunken Road and through the Piper cornfield, driving the Confederate skirmisher back to their main line.[38]
4. The 7th continues south to the Piper barn. When they crest the high ground north of the farm Colonel Thomas W. Hyde sees Confederates arrayed before them. He orders the regiment to move to the east into the orchard.[39]
5. Remnants from Wilcox's Brigade under Major Hilary A. Herbert and a mix from Ripley and Wright's brigades under Lt. Colonel Phil Cook arrive from the south.

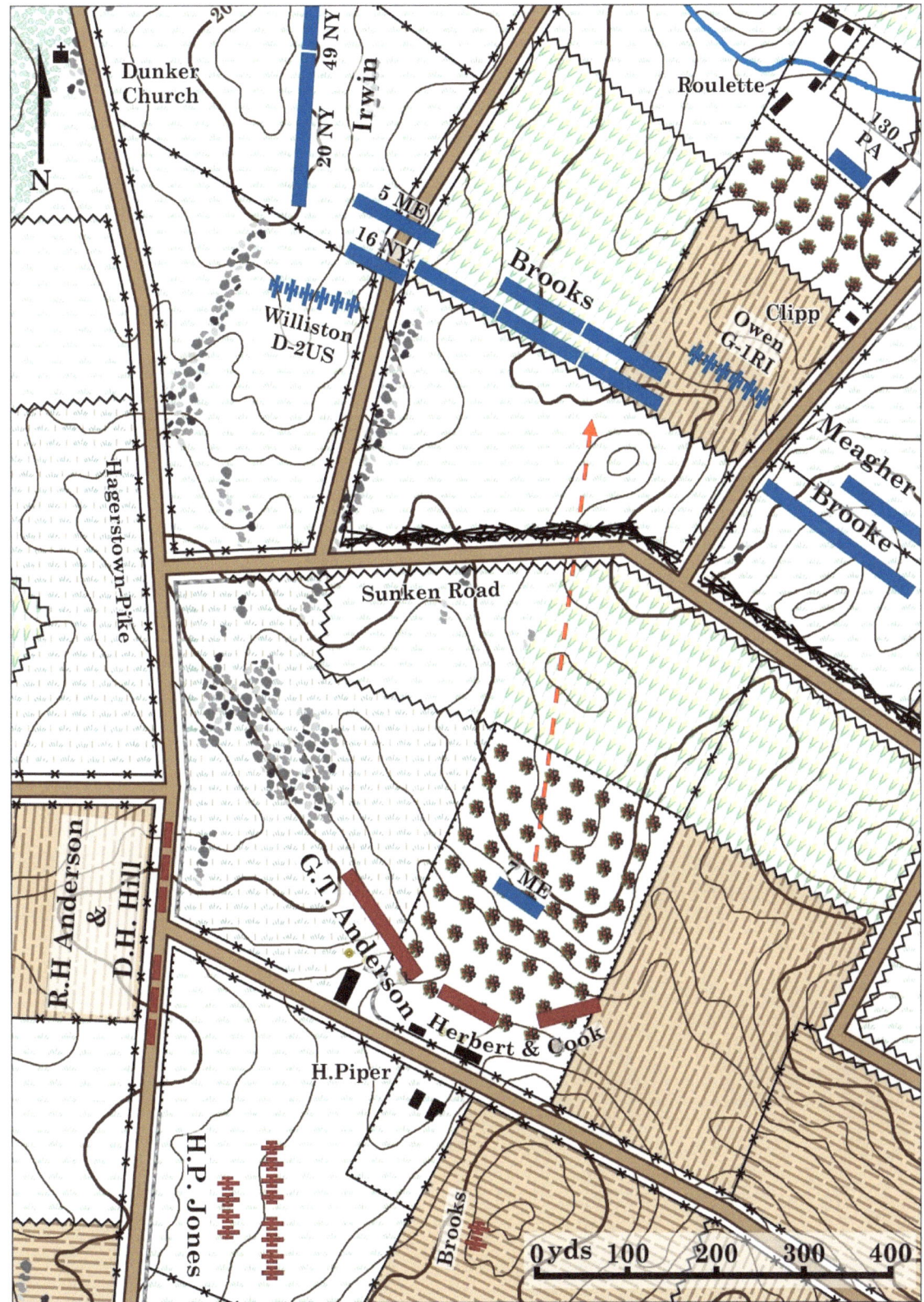

1. 5:15 p.m.
2. The Confederates advance on the 7th Maine in the Piper orchard from three sides.
3. The 7th retreats back to the Union lines.[40]

The Middle Bridge

The Middle Bridge over Antietam Creek provided the direct route into the town of Sharpsburg. General McClellan made his functional headquarters at the Philip Pry house north of the bridge. Numerous and plentiful artillery batteries occupied the high ground on the east side of the creek, where they commanded the approaches to the bridge and could even fire miles away at the battle raging to the north. Major General George Sykes and his division of predominantly United States Regular Army infantry supported the artillery at the bridge.

Confederate opposition was meager. Brigadier General David R. Jones' understrength division guarded the Boonsboro Pike from the bridge to Sharpsburg. Brigadier General Nathan G. Evan's independent brigade shored up his left flank north of the road. Both units held the high ground east of the town called Cemetery Hill.

The action around the Middle Bridge was comparatively quite as the morning battle raging to the north and later to the south at the Lower Bridge. Just before noon activity increased. Union cavalry crossed the bridge, drove back Confederate skirmishers, and established their own skirmish line. Union artillery advanced to the heights just below Cemetery Hill and began bombarding the rebels. General Sykes then sent forward several regiments to form a stronger skirmish line.

After 3 p.m. the Regulars advanced toward the town, with the Ninth Corps to the south moving forward with them, although the two units did not coordinate their attack. The Regulars forced the scant Confederates from the hill, and even staved off a limited counterattack. However, just as they were about to capture the hill, General Sykes recalled them to their starting point, feeling they had exceeded their orders and over-exposed themselves. The repulse of the Ninth Corps to the south removed any further options for an advance. Only intermittent skirmishing continued until nightfall.

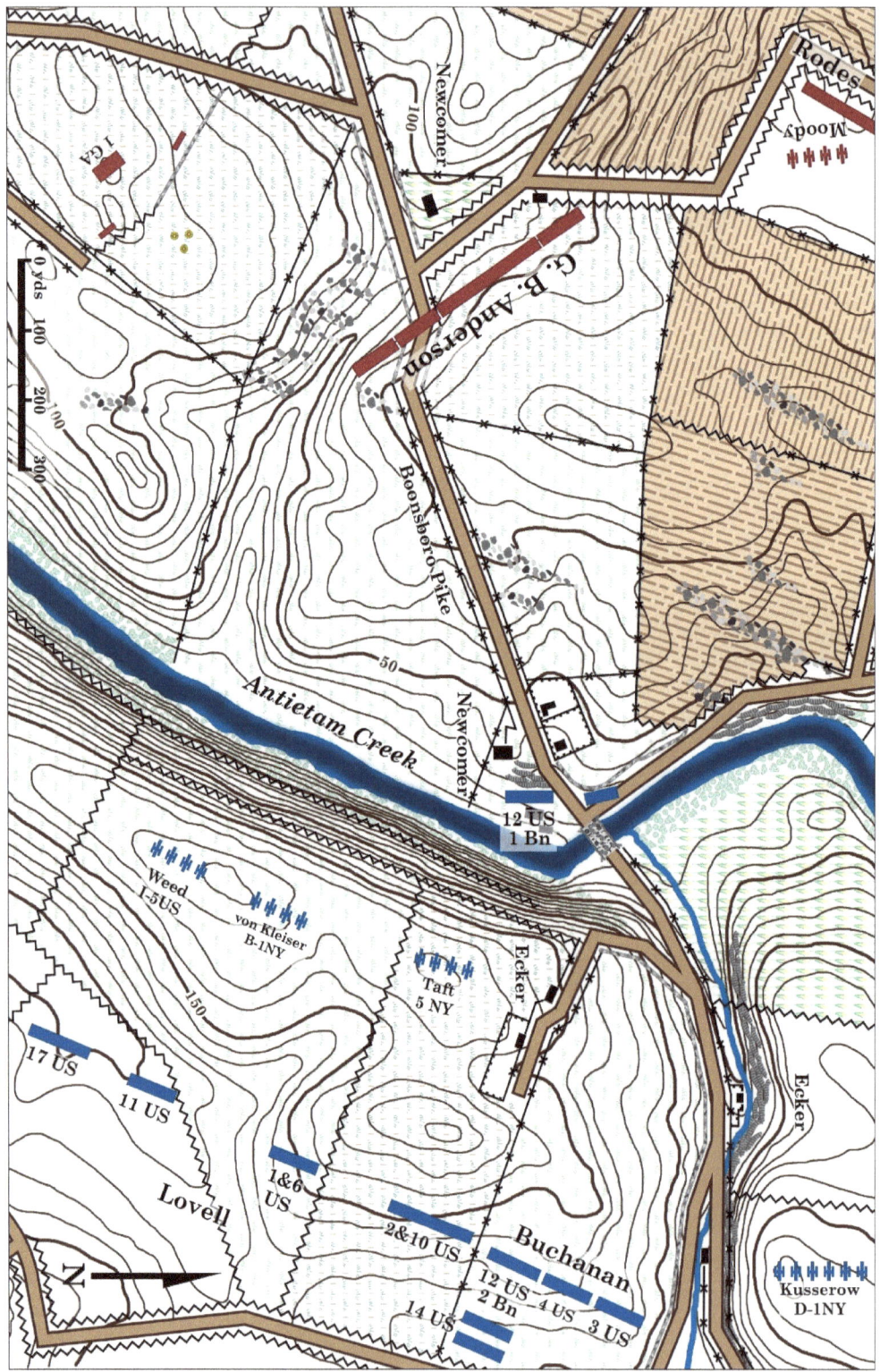

1. First light. Approximate 5-5:15 a.m.
2. The Middle Bridge at dawn.
3. The 1st Battalion, 12th United States holds the bridge and the Newcomer buildings.
4. George B. Anderson's brigade holds the reverse slope above the Newcomer farm. They will move north to the Sunken Road around 8:30 a.m.

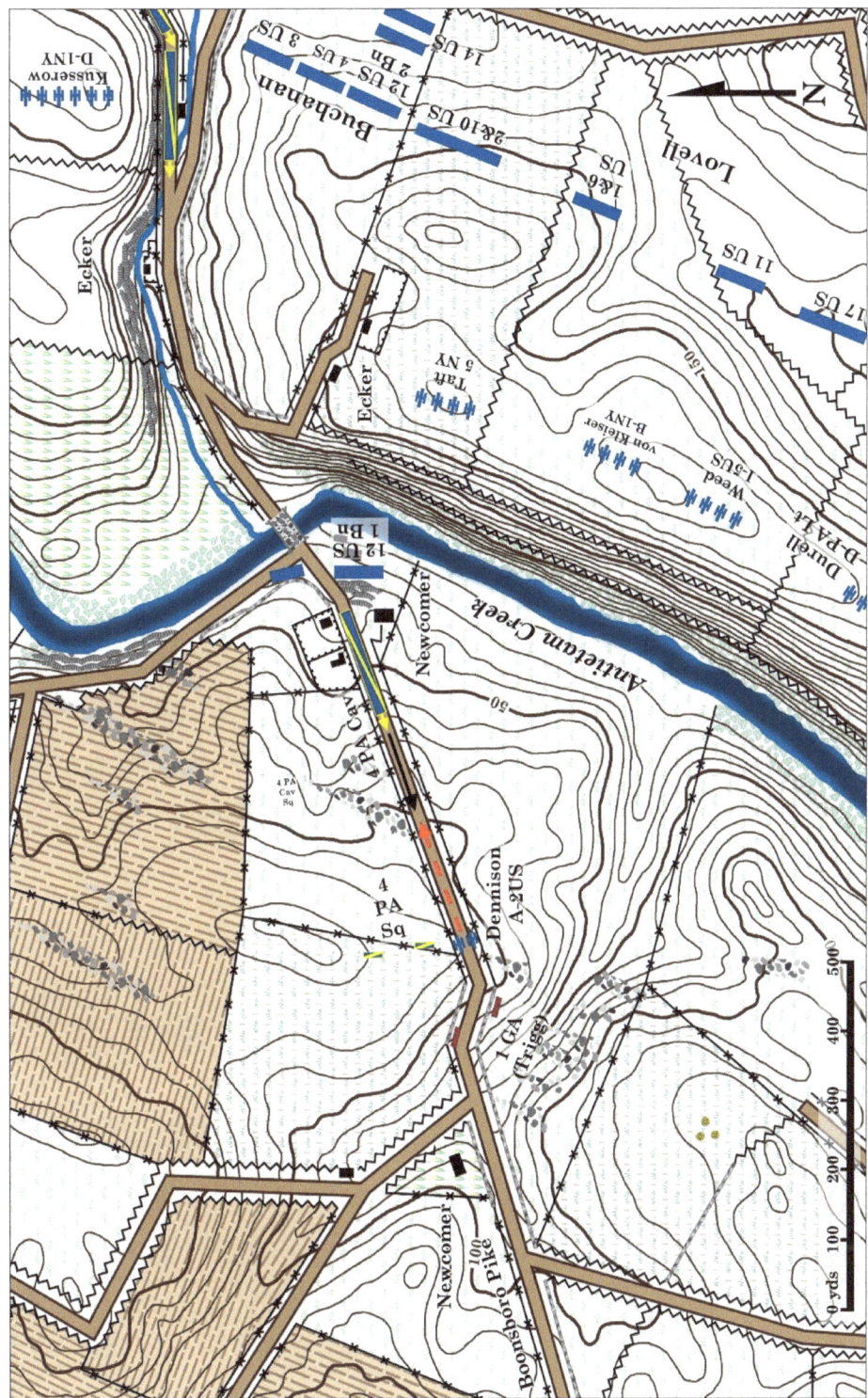

1. 11:45 a.m.
2. Dennison's section of Battery A, 2nd United States was moved forward to engage the Confederates, supported by a squadron from the 4th Pennsylvania Cavalry.[1]
3. Skirmishers from the 1st Georgia Regulars, left behind when their brigade moved north, engage Dennison and drive him off.[2]
4. The remainder of the 4th Pennsylvania Cavalry arrives and deploys behind the crest.[3]

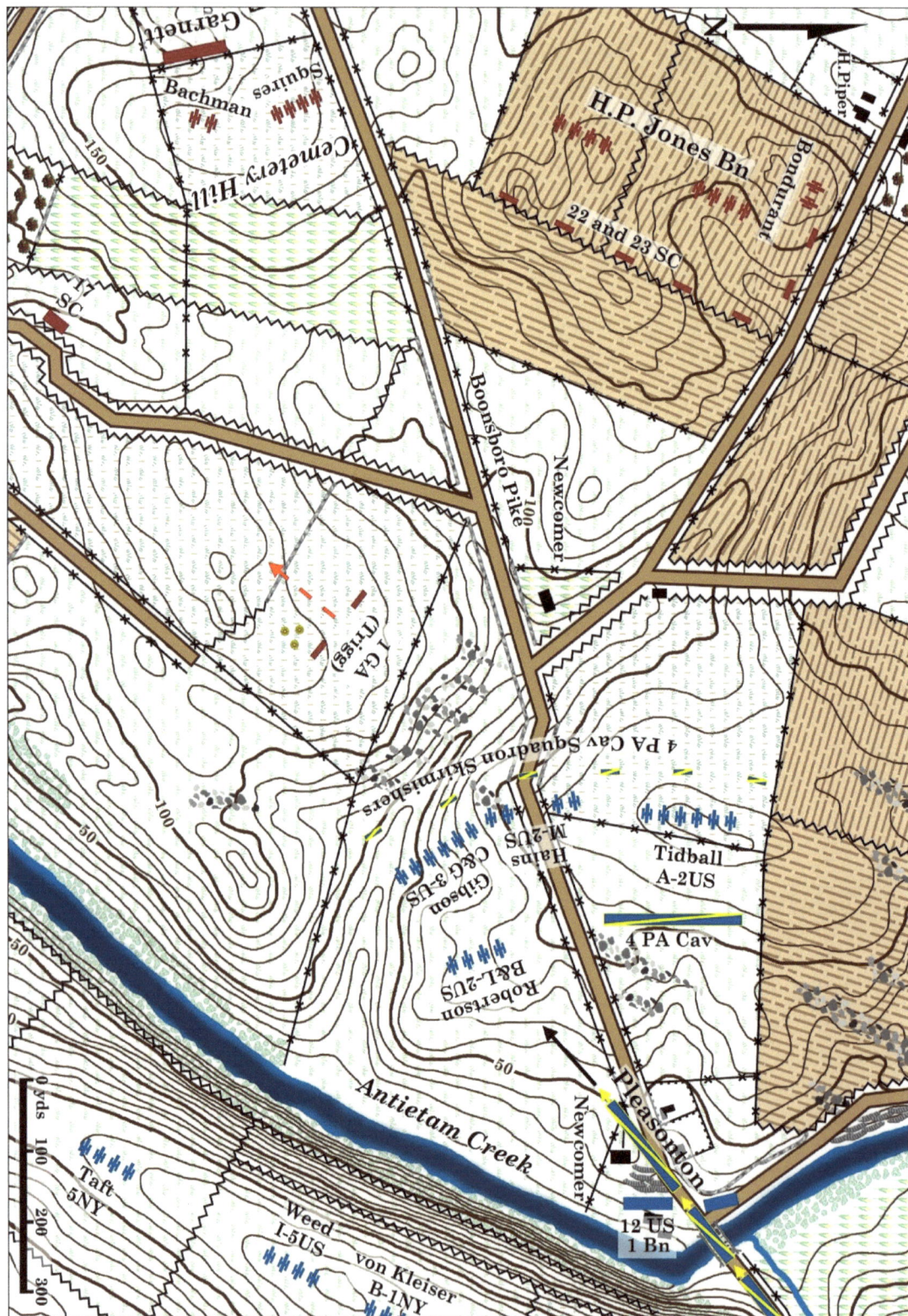

1. 12:15 p.m.
2. Pleasonton arrives with the bulk of the Union cavalry.
3. The Union horse artillery deploys on the high ground above the Newcomer buildings.
4. The 1st Georgia Regulars skirmishers are forced to fall back.

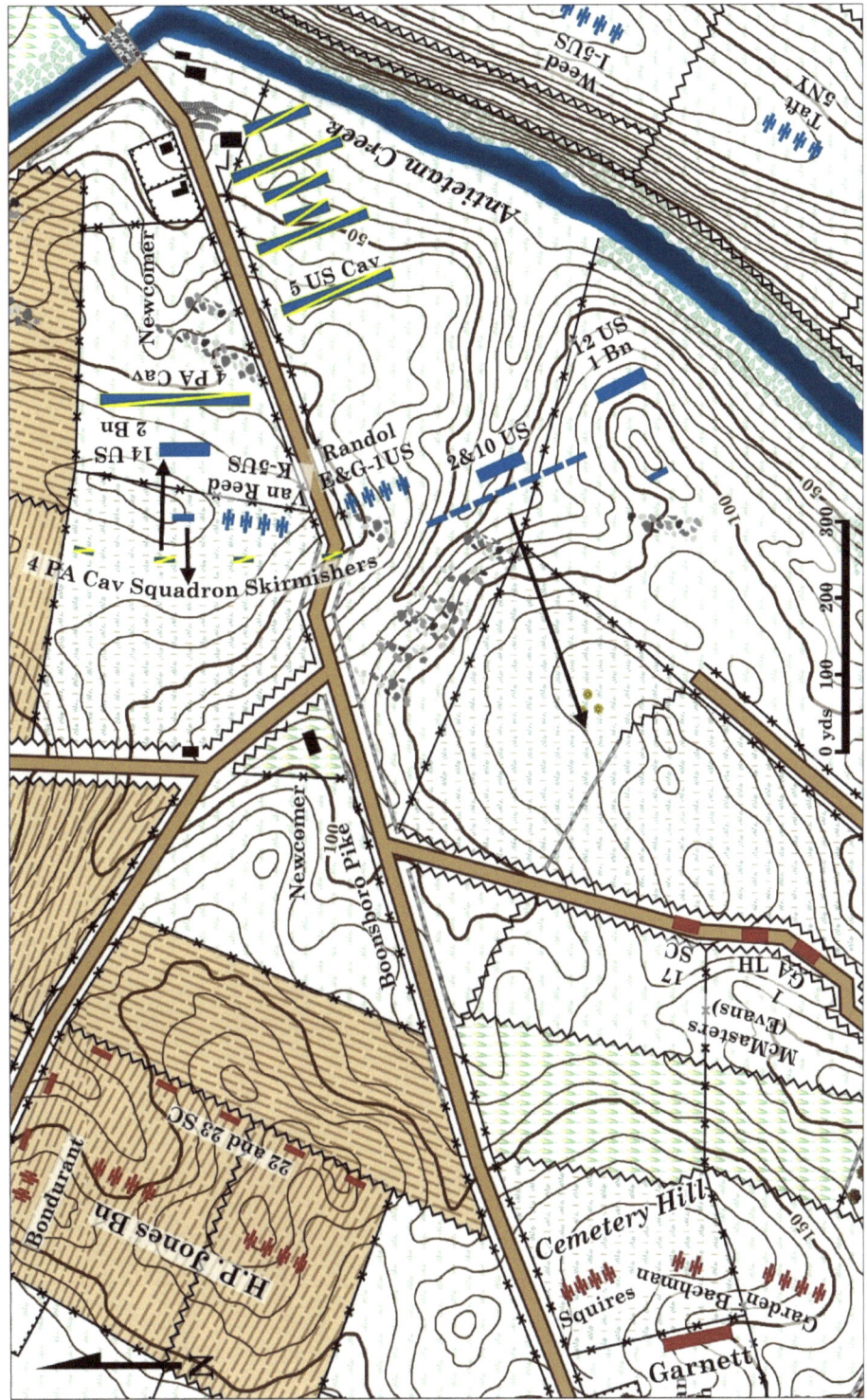

1. 12:30 p.m.
2. Pleasonton orders the 1st Battalion, 12th United States, to his aid. Sykes sends the 2nd & 10th United States to support.[4]
3. Sykes commands Captain Hiram Dryer to take charge of all US Regulars on the west side of the creek with orders to support the batteries, and "dislodge the enemy from certain haystacks."
4. The 2nd Battalion, 14th United States relieves the 4th Pennsylvania Cavalry skirmishers.[5]

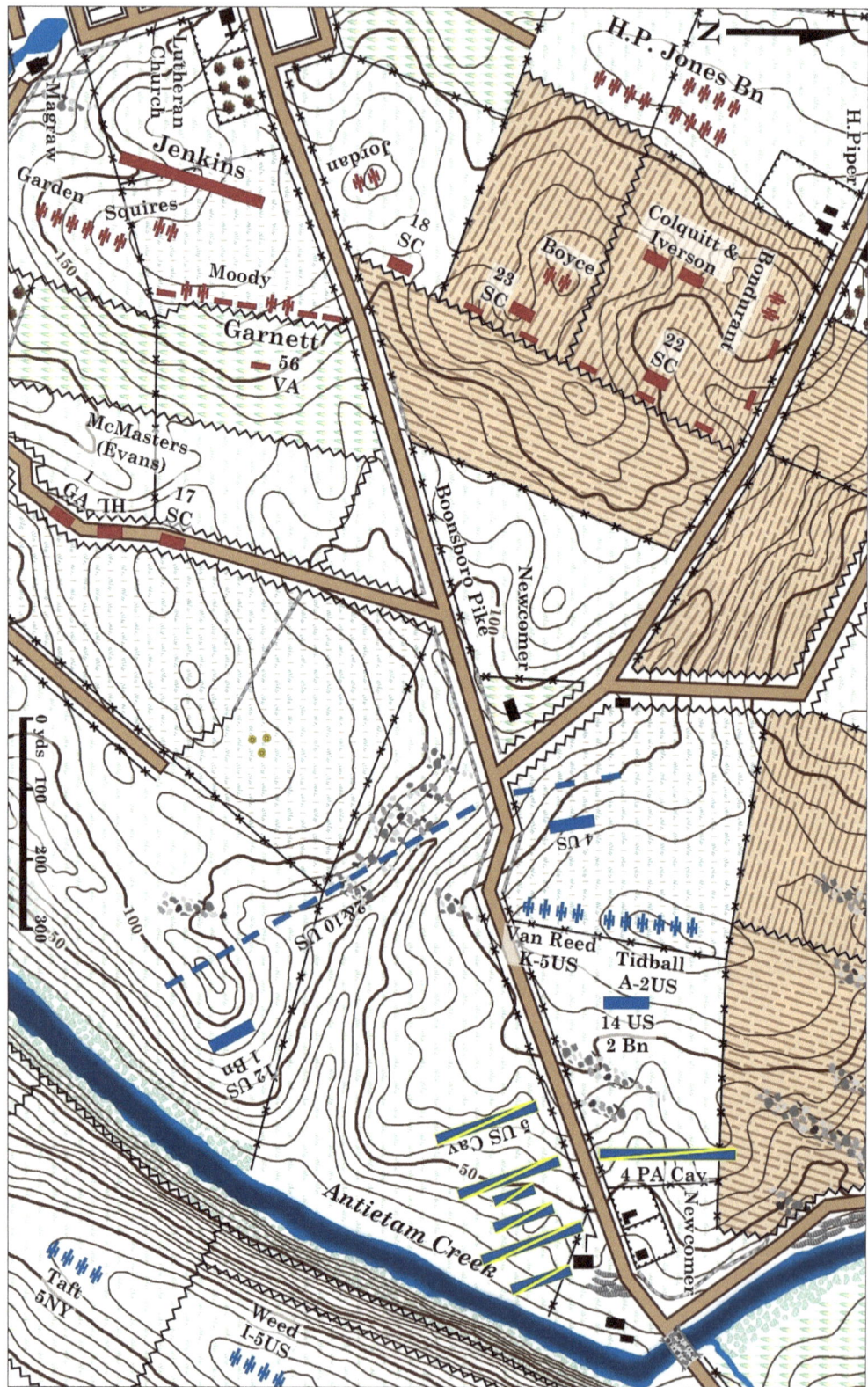

1. 1:00 p.m.
2. The US Regulars take up a position guarding the Union artillery on the west side of Antietam Creek.
3. Pleasonton's cavalry take shelter in the low ground near the Middle Bridge.

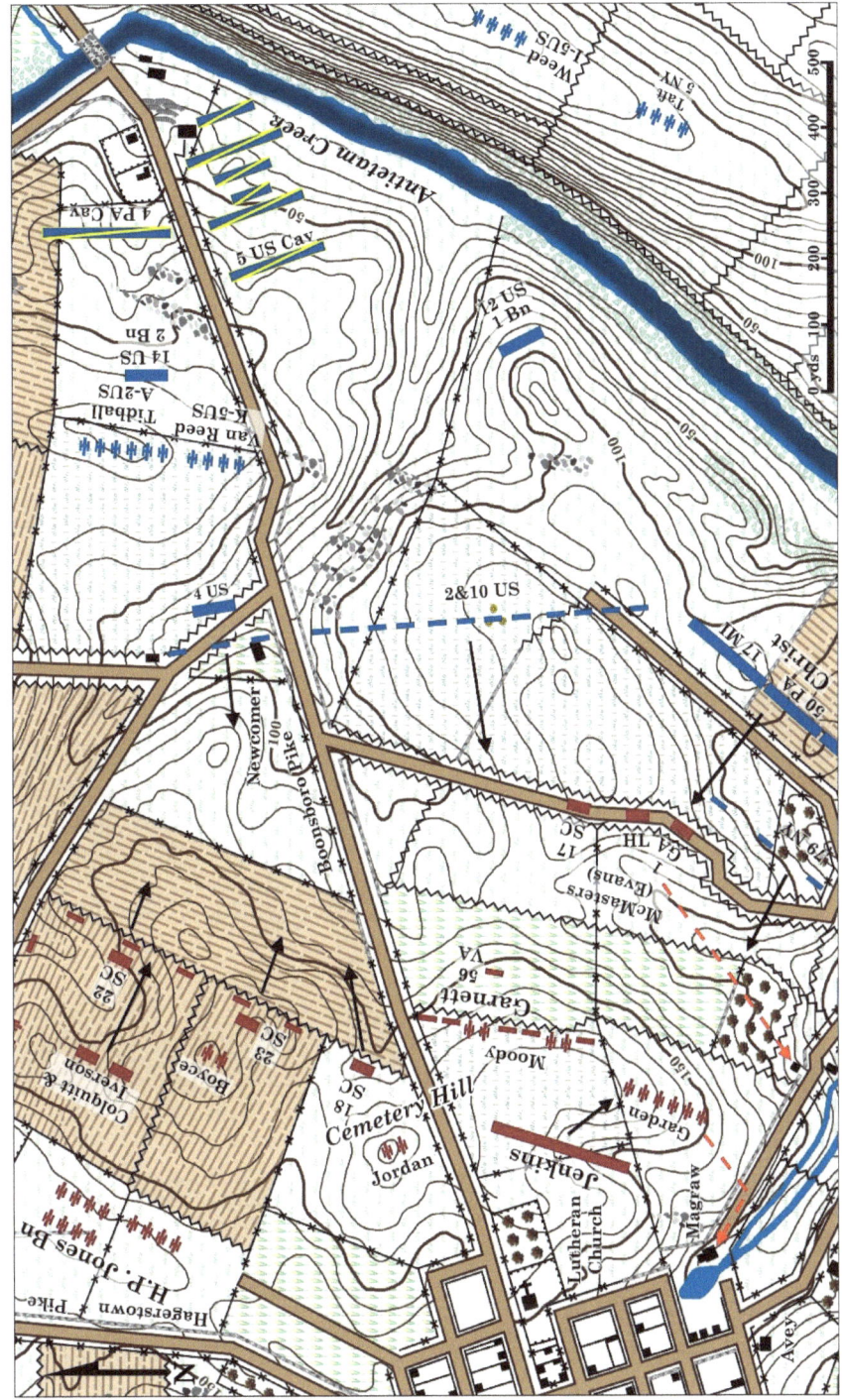

1. 3:30 to 3:45 p.m.
2. Captain Dryer orders the skirmish line forward.[6]
3. Christ's Ninth Corps brigade advanced through the Sherrick farm and toward the southern slope of Cemetery Hill. The two advances are complimentary, but not coordinated.
4. The 18th, 22nd and 23rd South Carolina from Evans Brigade, and stragglers under Colquitt and Iverson, advance to meet them.[7]
5. Christ's advance forces the skirmishers in the Sherrick lane and Garden's battery to retreat.
6. Jenkin's Brigade moves forward to engage Christ.

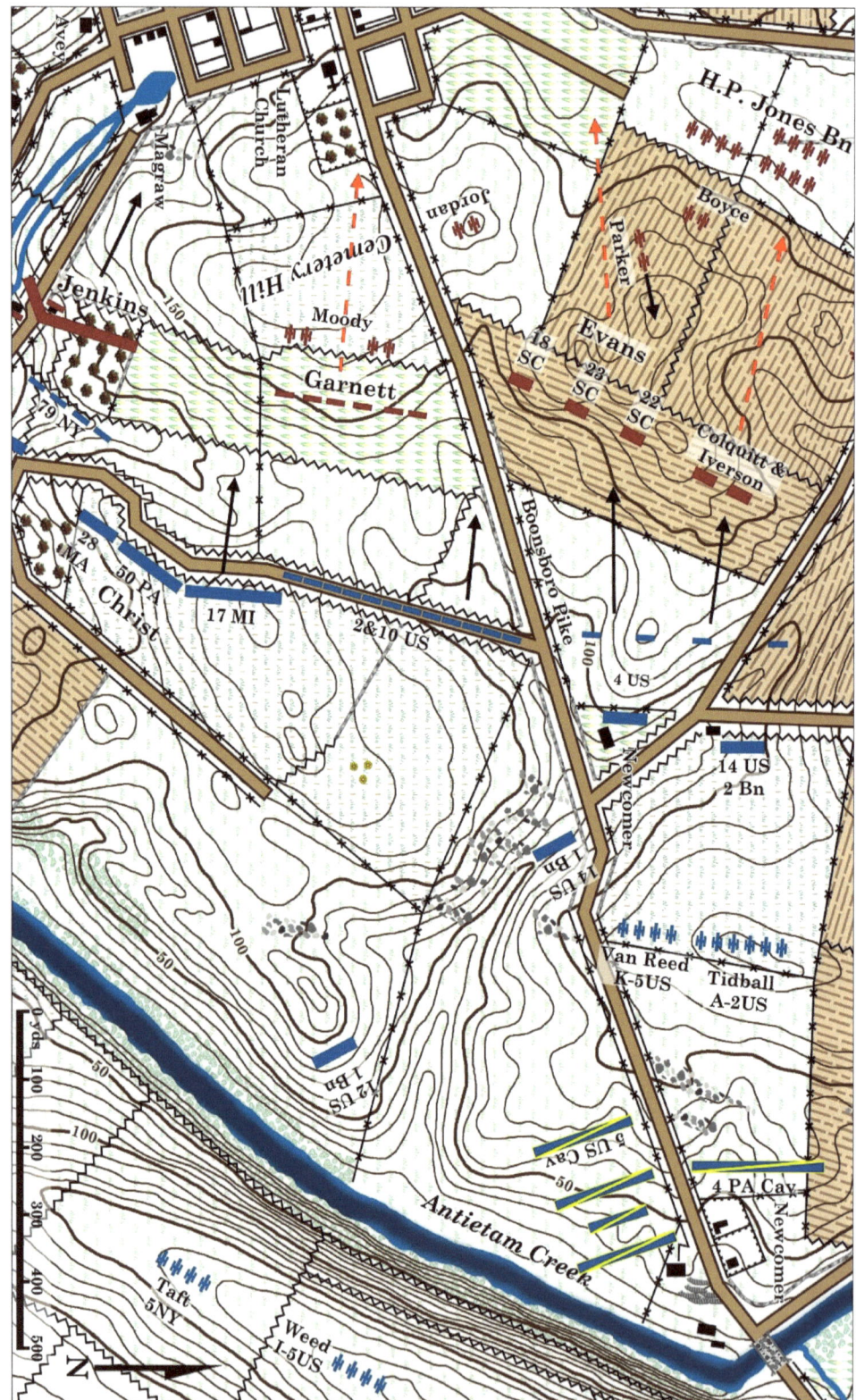

1. 3:50 p.m.
2. The Union line continues its advance, forcing Evans, Iverson/Colquitt, and Garnett to retreat.[8]
3. Jenkins falls back over the slope of Cemetery Hill in the face of pressure from the front and south.

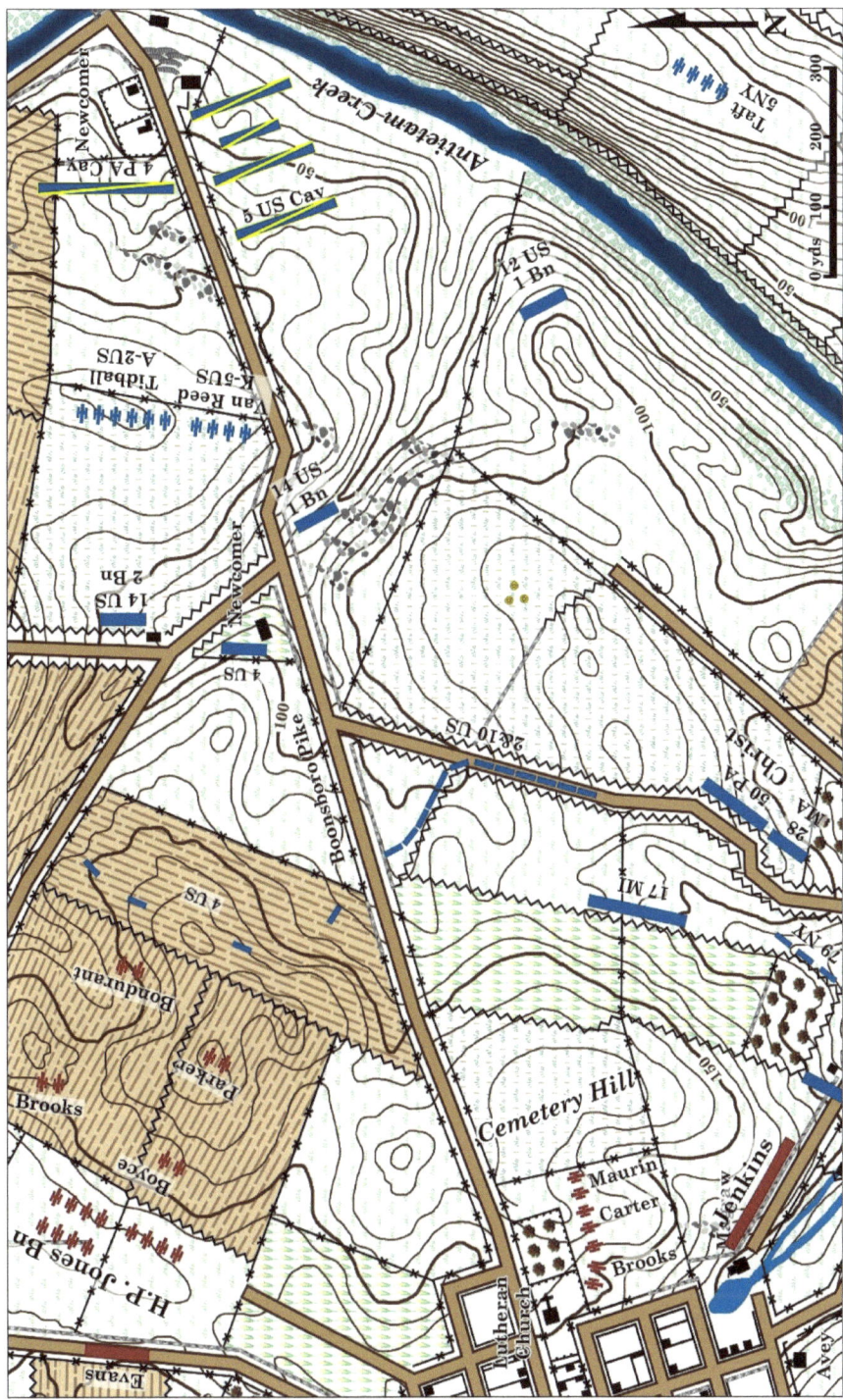

1. 4:15 p.m.
2. Sykes orders Captain Dryer to stop his pursuit of the retreating Confederates just as Cemetery Hill is clear of infantry. Dryer protests, but obeys and the advance is halted.[9]
3. Several Confederate batteries arrive to confront the Union advance. The batteries of Brooks, Carter, and Maurin deploy on the south reverse slope of Cemetery Hill and bolters Jenkins.[10]
4. Jenkins forms facing south with his left flank protected by a spur of Cemetery Hill.
5. Christ's brigade halts and does not proceed further due to artillery fire from farther south striking his flank, and difficulty resupplying ammunition.[11]

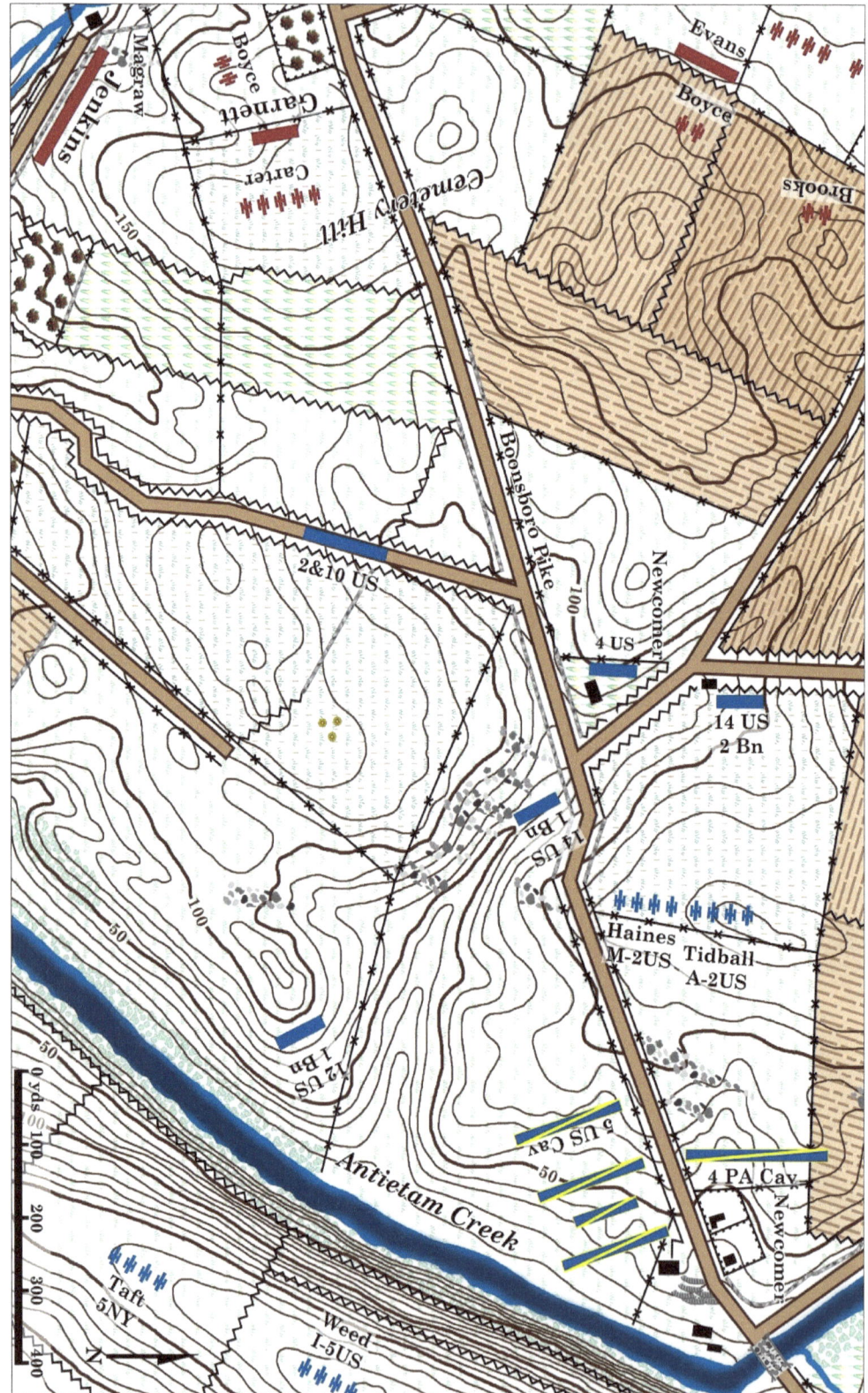

1. 6:00 p.m.
2. The Confederates end the day east of Sharpsburg with little infantry available. Only a few batteries support them against the Union line protecting the Middle Bridge.

Burnside's Bridge

The fight for the Lower Bridge, soon to be named Burnside's Bridge after the Union area commander, was marked by inadequate reconnaissance, misunderstood orders, and at times, poor performance. The Union command structure was hampered by unnecessary duplication. Major General Ambrose E. Burnside had been a wing commander in charge of two army corps up until a few days ago. McClellan had rescinded that organization on the eve of the battle, but Burnside remained under the impression that he was still a wing commander. Therefore Brigadier General Jacob D. Cox retained operational control of the Ninth Corps. Two levels of corps command interfered with the quick and easy receipt and implementation of orders from army headquarters.

The 17th dawned with the corps concentrated around the bridge, but the first attacks didn't begin until about 10 a.m. They were individual and not coordinated on a large scale. Beginning with a single regiment, brigade after brigade were thrown at the bridge without success. Opposing them were two regiments from Brigadier General Robert Toombs' Brigade, commanded by Colonel Henry L. Benning, another detached regiment, and artillery support. Only when the plentiful Union artillery began a concentration of fire, and regiments provided adequate rifle covering fire, were the attackers able to dash across the bridge and drive back Benning's men from the heights defending the crossing.

While this fight was occurring Brigadier General Isaac P. Rodman marched south along the Antietam in search of fords to cross and outflank the defenders. The ford recommended by army headquarters was unsuitable for an entire division, so Rodman continued downstream until he found Snavely's Ford. There, he crossed his division over, drove back the thin line of defenders, and marched north to link up with the rest of corps on the west side of the creek.

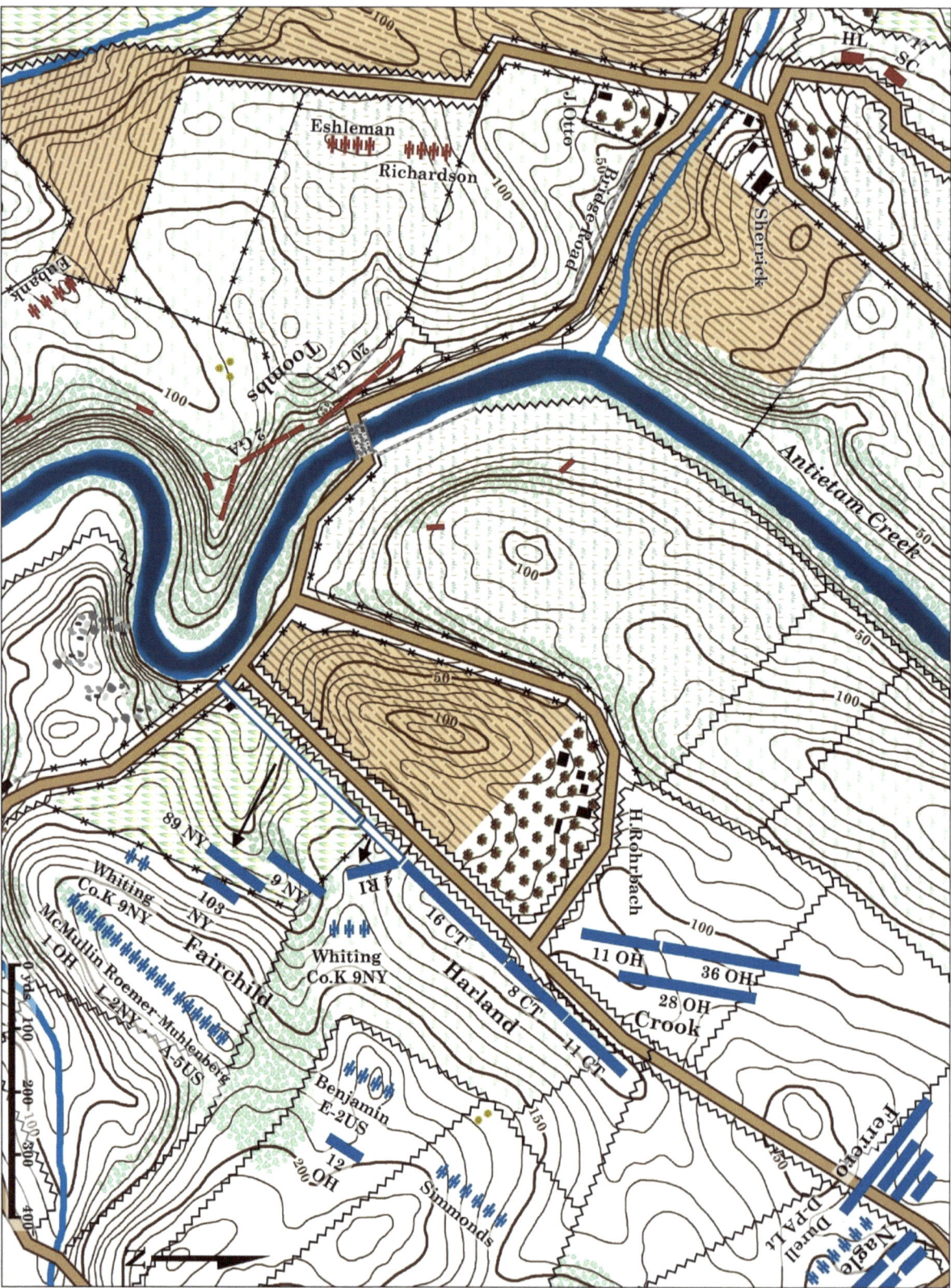

1. First light. Approximate 5-5:15 a.m.
2. As the day brightens Eubank's Battery realizes they have an enfilading flank shot on Fairchild's brigade. Fairchild will move his brigade to cover after taking several casualties.[1]

1. 10:00 a.m.
2. The 11th Connecticut attempts to carry the bridge. Companies A and B try to ford the creek below the bridge, but are driven back. Captain John D. Griswold is killed leading the attempt.[2]
3. After the 11th Connecticut's Colonel Henry W. Kingsbury is mortally wounded, the regiment falls back.[3]
4. Crook's brigade advances toward the bridge.

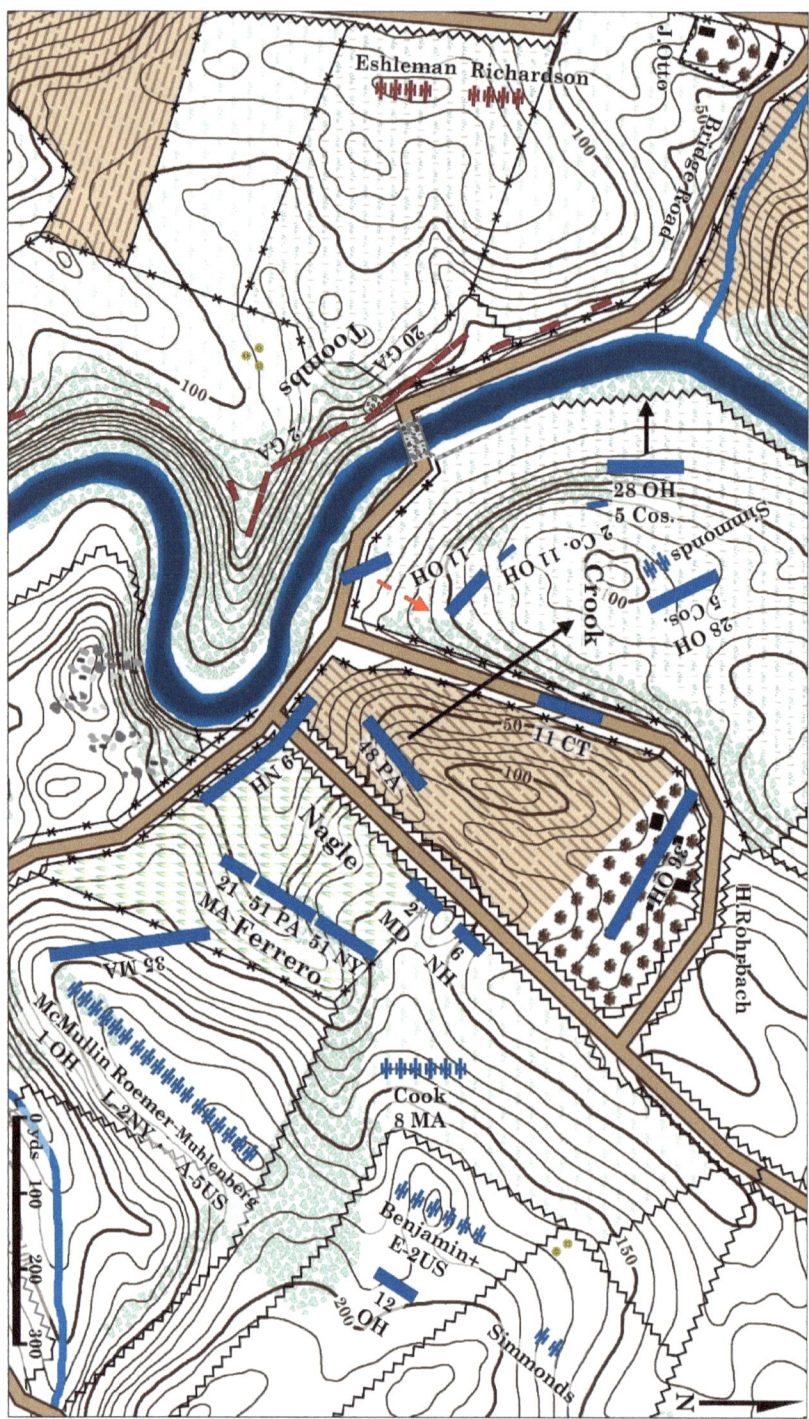

1. 11:00 a.m.
2. Crook makes a disjointed assault on the Lower Bridge.
3. The 11th Ohio splits and part of the regiment makes it to the creek. Lt. Colonel Augustus H. Coleman is mortally wounded and the regiment falls back.[4]
4. Half of the 28th Ohio advances towards the bridge as skirmishers and are repulsed. Crook leads the other half toward the bridge, but fire causes them to drift to the north towards a bend in the creek. They remain there and engage the skirmishers across the creek.[5]
5. The 48th Pennsylvania from Nagle's brigade marches to the high ground and deploys into line.[6]

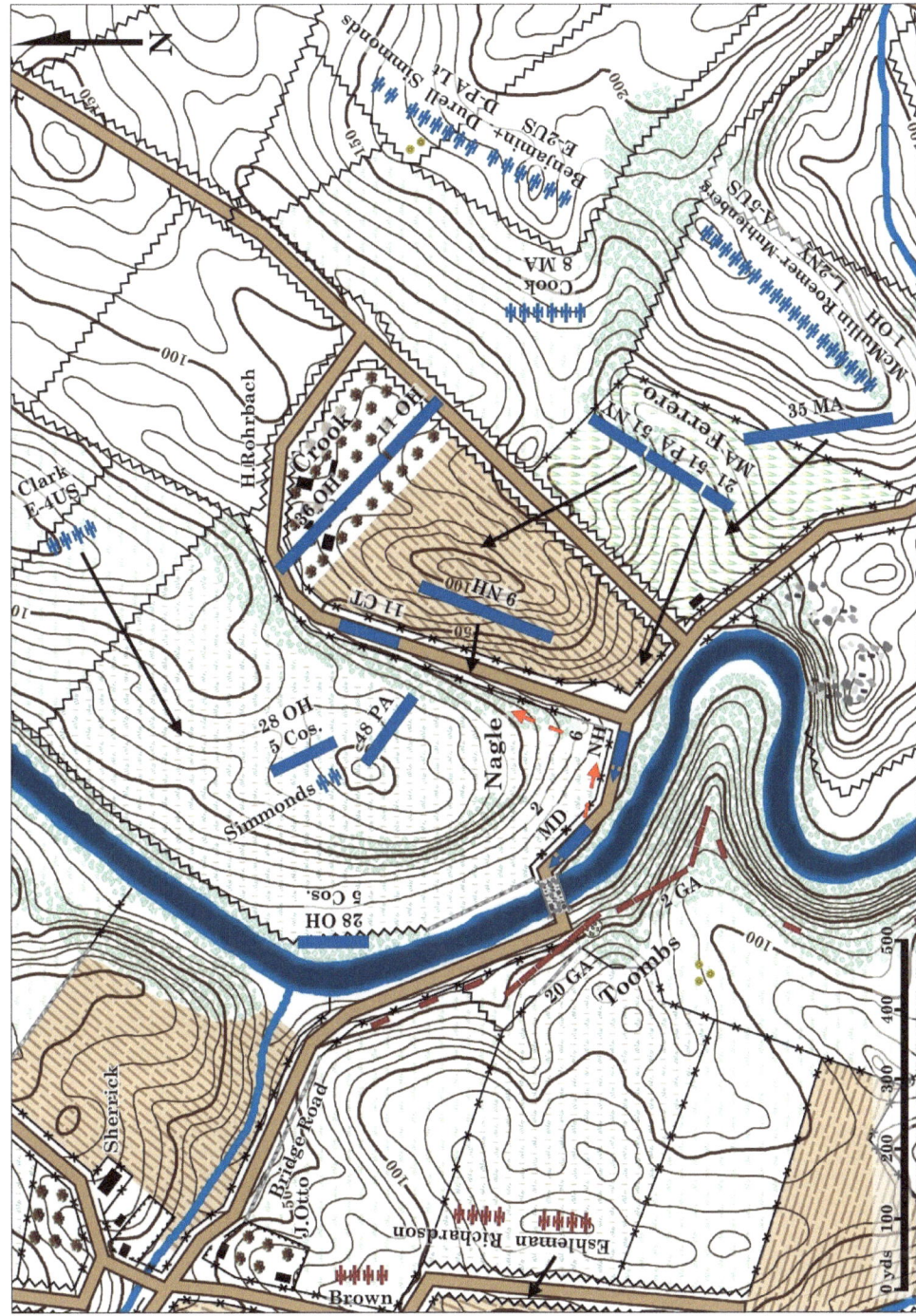

1. 11:30 a.m.
2. All artillery concentrates on the Confederates above the bridge.
3. Nagle's brigade charges the bridge. The 48th Pennsylvania provides covering fire from the heights above the bridge.[7]
4. The 2nd Maryland and 6th New Hampshire attempt to take the short route along the road. They are met with a murderous fire and forced to retreat. However, many men remain along the thick fence bordering the road and fire at the Confederates above them.[8]
5. Sturgis orders Ferrero's brigade forward.[9]
6. Battery E, 4th United States arrives from the Middle Bridge.

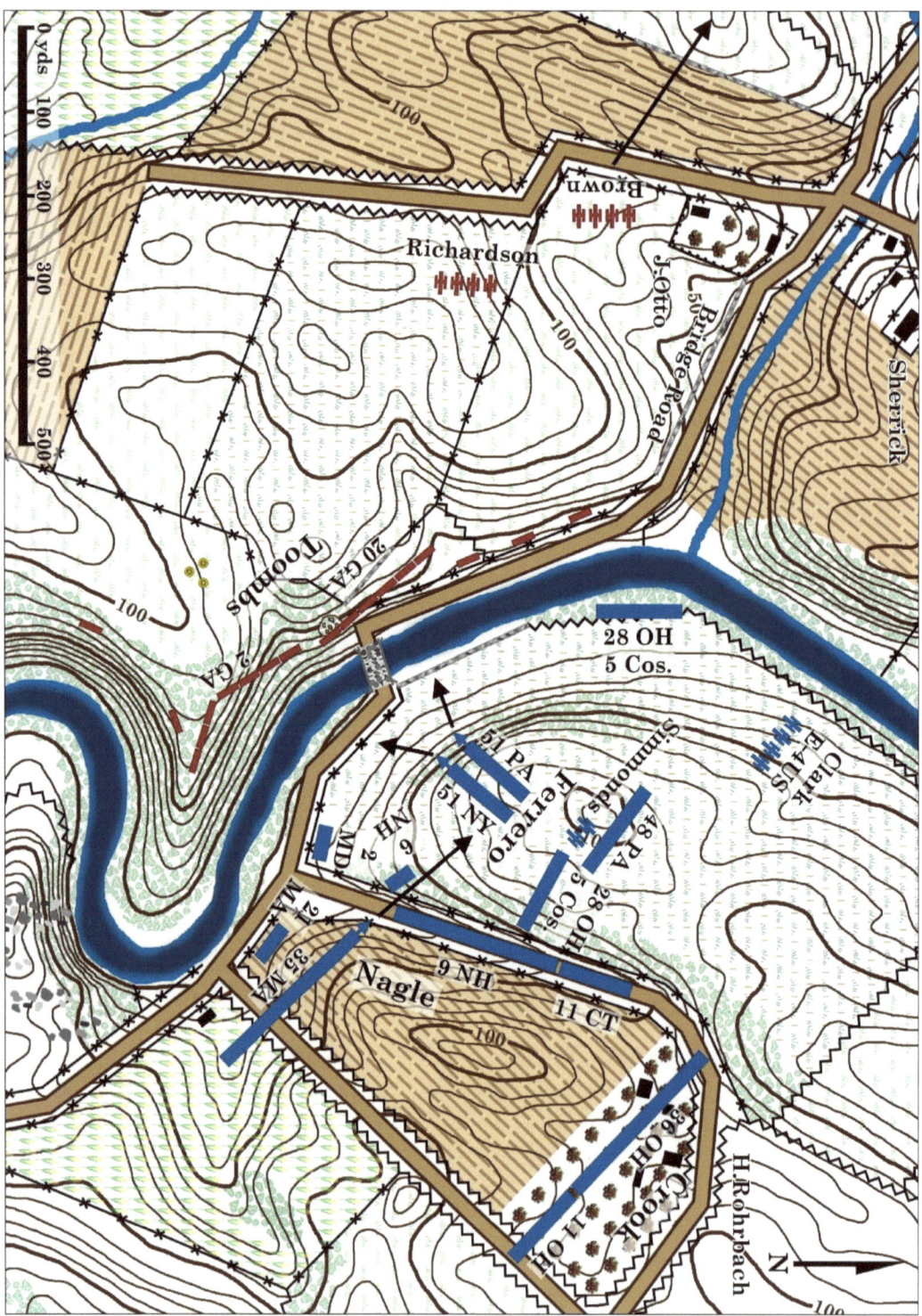

1. 12:30 p.m.
2. The 51st Pennsylvania and 51st New York rush the bridge in column. Taking heavy fire, they seek cover along the fences on either side of the bridge.[10]
3. The 35th Massachusetts arrives and forms another column behind them.[11]
4. The Confederates continue to endure a heavy fire from the concentrated artillery bombardment and the Union infantry gathering above and below the bridge.

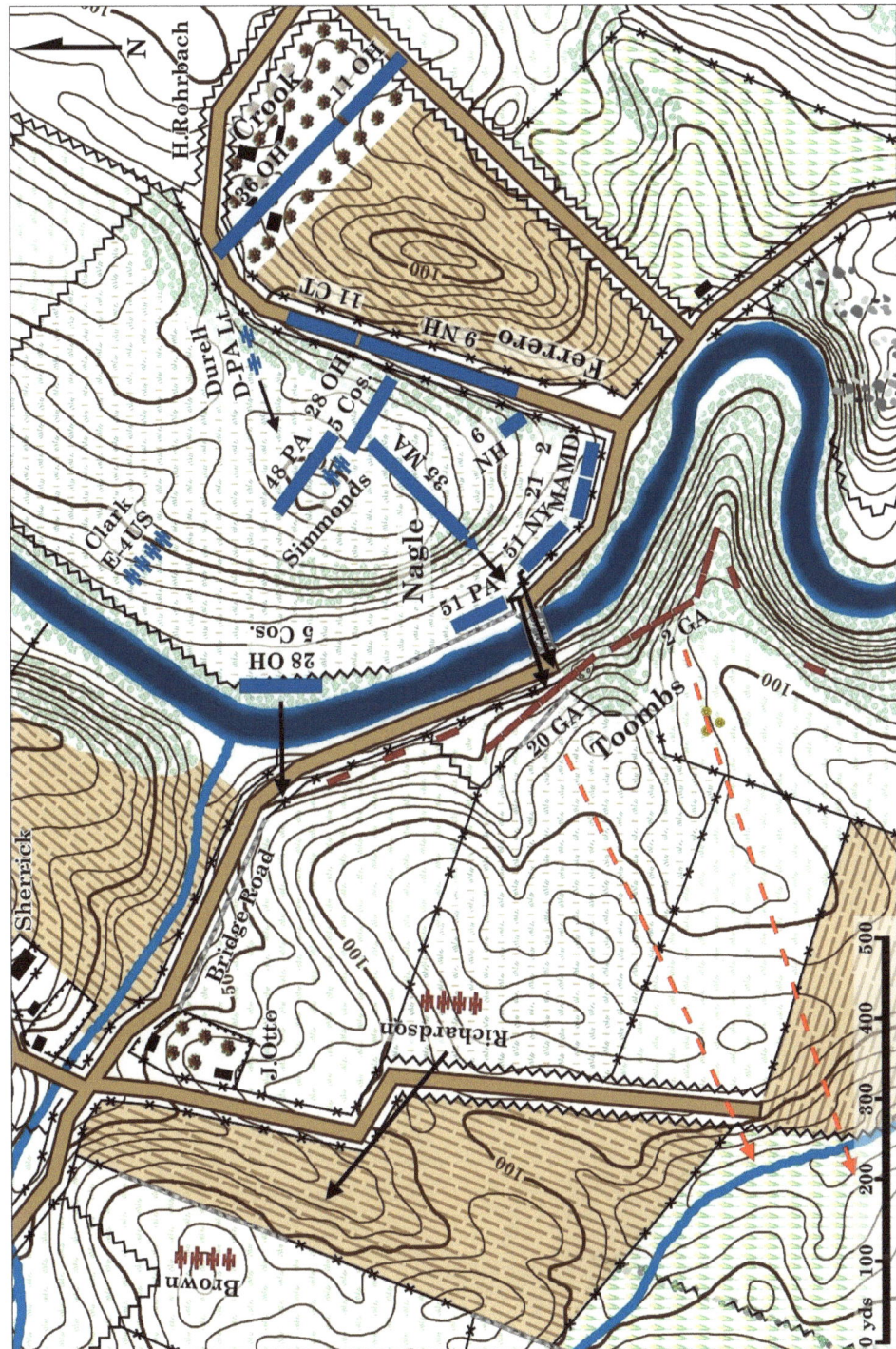

1. 1:00 p.m.
2. The Confederates begin to fall back in small groups of 2-3 men.
3. The commanders of the 51st Pennsylvania and 51st New York see the Confederates beginning to waver, and charge across the bridge, the colors alongside each other.[12]
4. The 35th Massachusetts follows closely behind them.
5. Crook crosses over the creek with the five companies of the 28th Ohio at a ford.[13]
6. The overwhelming pressure forces Toombs' Brigade and the remainder of his supporting artillery to retreat.

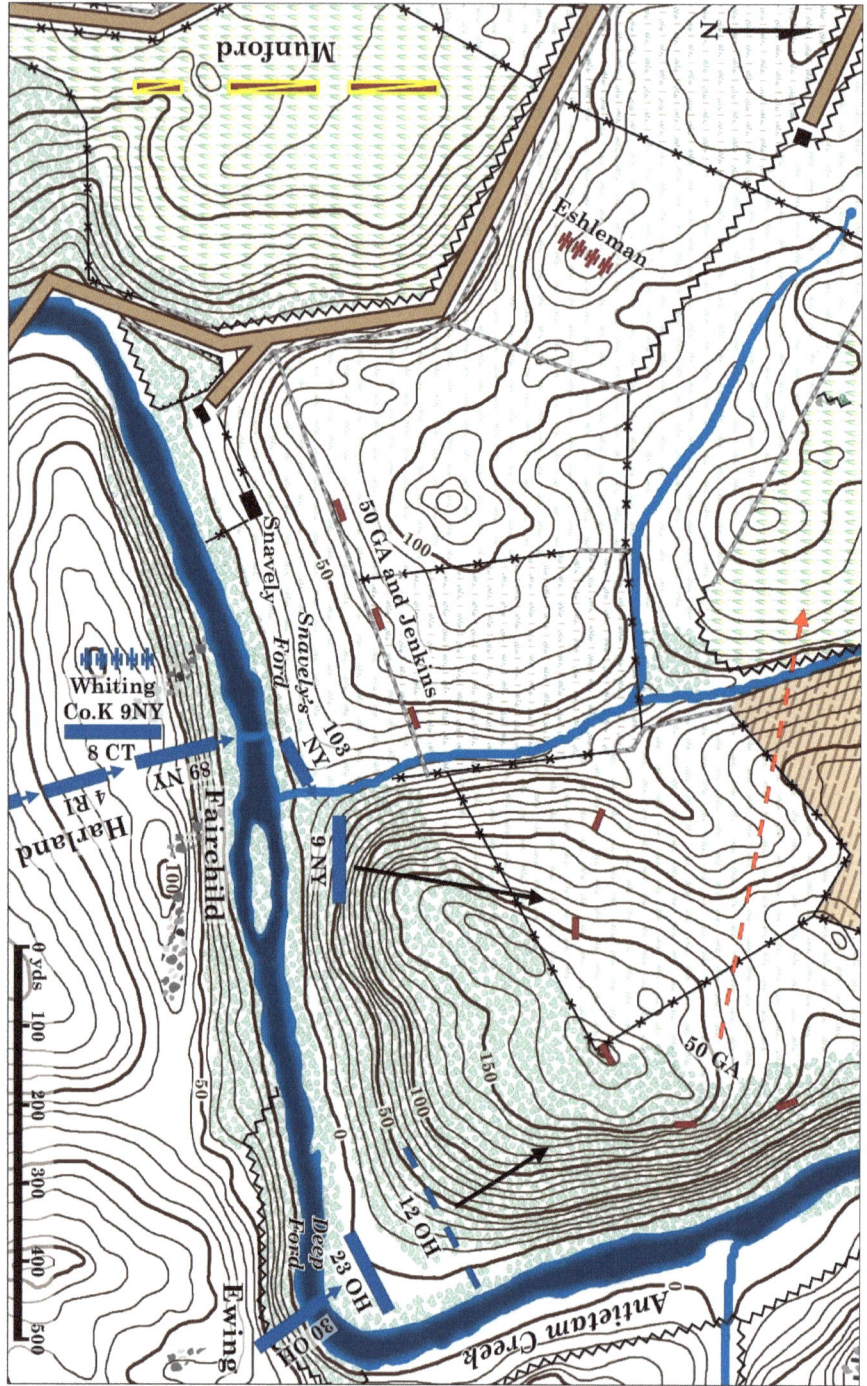

1. 1:00 p.m.
2. Ninth Corps commander Jacob D. Cox orders Rodman to take is division downstream and cross at the ford there. Rodman finds the first ford, Deep Ford, impractical for the entire division, and continues on to Snavely Ford.[14]
3. Ewing's brigade crosses at Deep Ford and drives back the 50th Georgia skirmishers.
4. Fairchild and Harland cross at Snavely's Ford. The 9th New York assists in driving back the 50th Georgia.[15]
5. Fairchild and Harland drive back the skirmishers on the heights above Snavely's Ford.[16]
6. The division marches north to link up with the rest of the Ninth Corps.

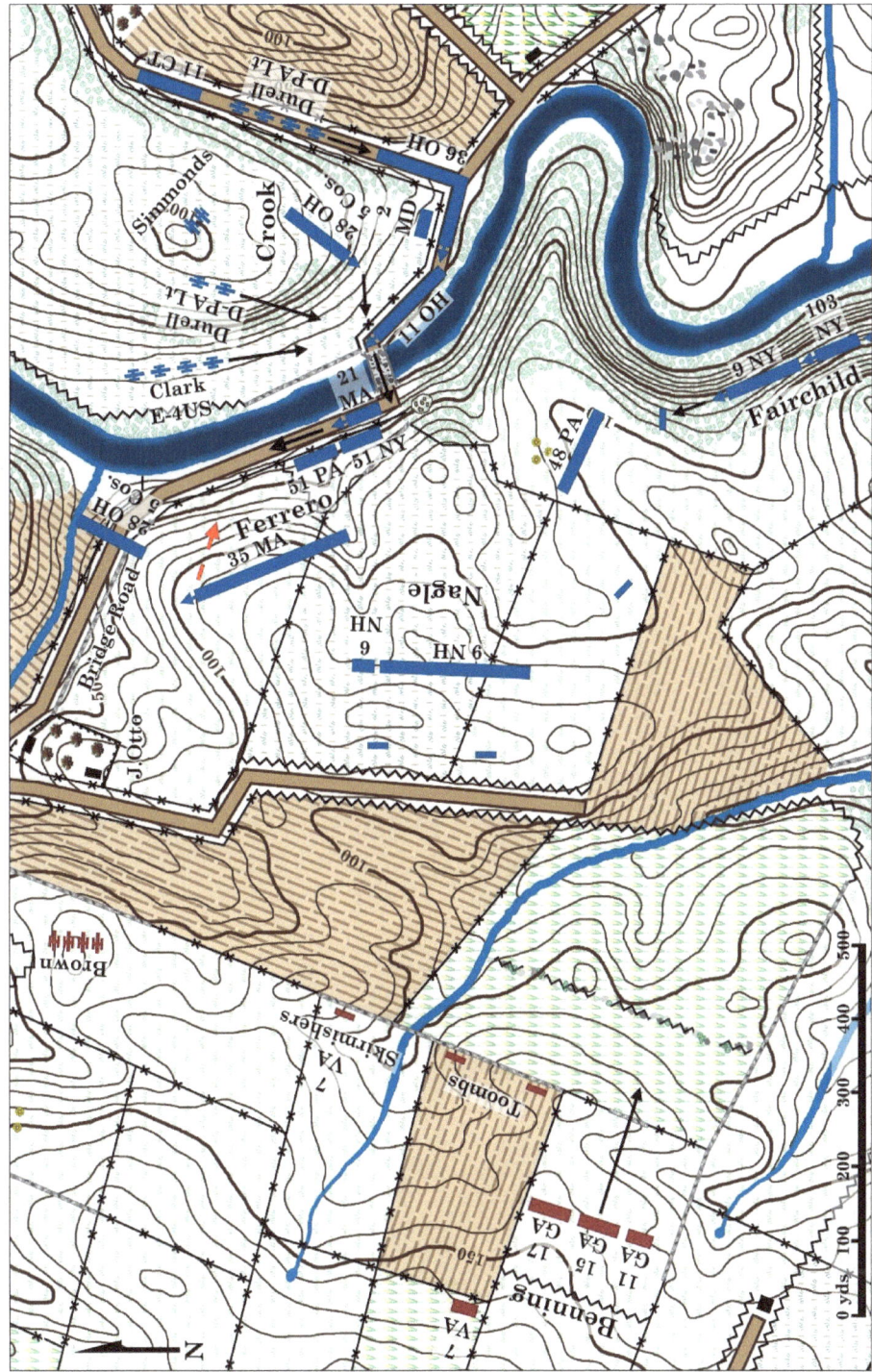

1. 1:15 p.m.
2. Once the bridge is secure the Union rapidly push more units across and deploy on the heights above.
3. The 35th Massachusetts advances up the high ground first in line, then in column because of the difficult terrain. As it crests the heights it receives fire from Brown's Battery and falls back below the crest and deploys.[17]
4. Nagle deploys on the high ground south of the bridge and establishes contact with Rodman, who is marching up from the fords to the south.

Ninth Corp Advance and Hill's Counterattack

After capturing the Lower Bridge, General Cox went to work organizing his command for a renewed advance. Brigadier General Orlando B. Willcox's division formed the right, General Rodman's the left, and the divisions of Brigadier General Samuel D. Sturgis and Colonel Eliakim P. Scammon in reserve. Cox's orders, from McClellan through Burnside, were to gain the heights to the south of Sharpsburg, along which ran the road to Harper's Ferry, capture the town, and if possible, continue and cut the rebel army off from Boteler's Ford beyond. This would trap Lee and his army on the east side of the Potomac River.

Opposing him were the scattered remnants of Toombs' and D. R. Jones' divisions. However, help was on the way. Major General Ambrose P. Hill's Light Division had arrived after a seventeen mile march from Harper's Ferry. Lee directed Hill to defend the heights along the Harper's Ferry Road south of town. As the Ninth Corps moved forward, Hill began arriving and taking position.

Willcox's division drove the limited Confederate defenders from the southern slope of Cemetery Hill just east of the town. However, once there they halted. Limited ammunition and difficulties resupplying it, coupled with increasing artillery fire into their left flank, stalled the division's advance. To Willcox's left, Cox ordered Colonel Harrison S. Fairchild to charge his brigade against the Confederate artillery on the high ground across the valley. In a brilliant charge, Fairchild drove away two rebel brigades and gained the heights. However, Rodman's other brigade under Colonel Edward Harland ran headlong into A. P. Hill's division. Harland's untried regiments crumbled before Hill's veterans, and the flank of the entire corps collapsed. Rodman was mortally wounded. Fairchild was forced to retreat, and a brigade sent forward to help Harland was overwhelmed. Cox recalled Willcox and Fairchild.

The day ended with the two lines engaging each other across a deep valley south of the John Otto farm until and after sunset. Against overwhelming odds, Lee's Army of Northern Virginia had managed to survive the Union onslaught.

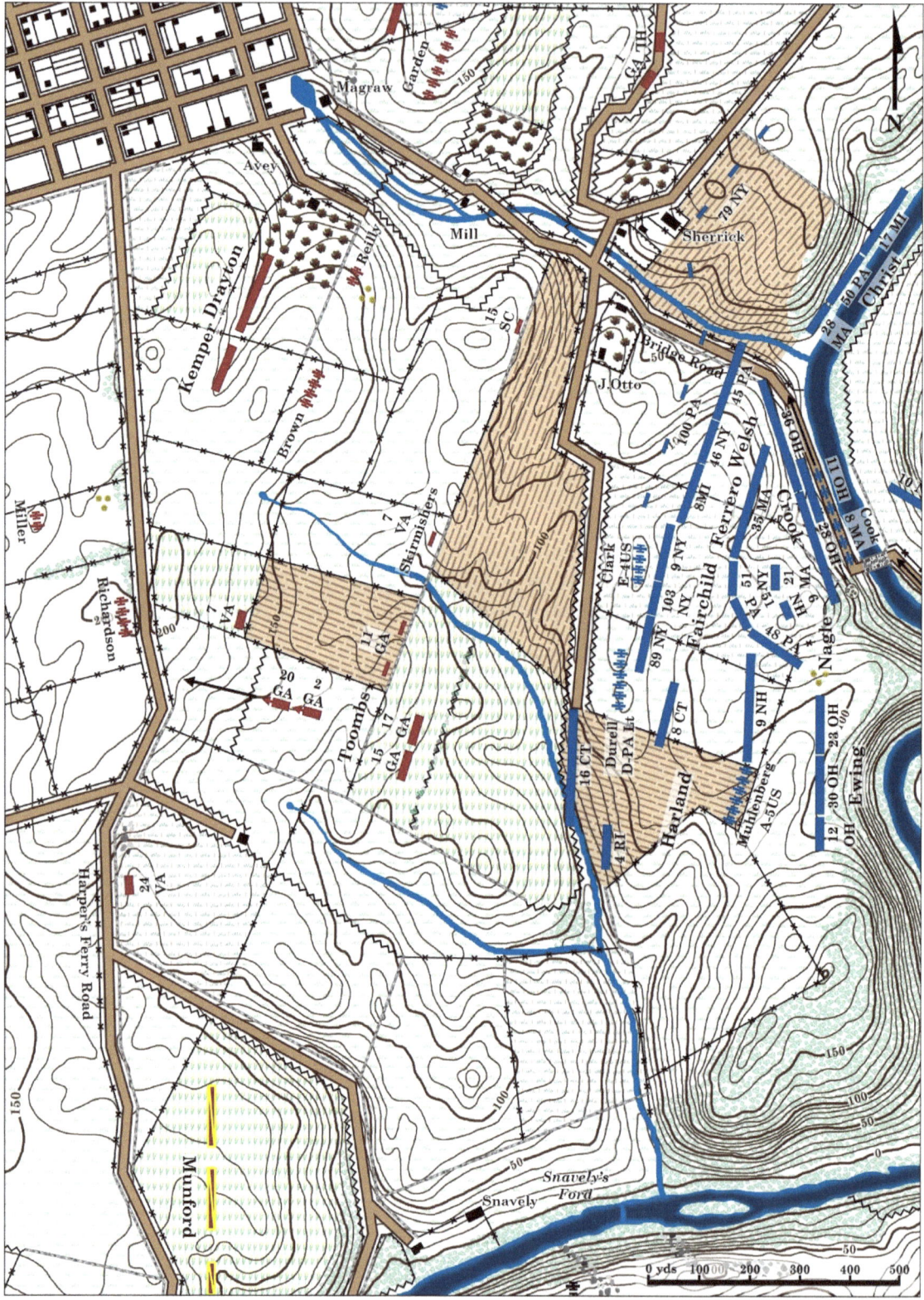

1. 3:00 p.m.
2. Initial disposition of the Ninth Corps after crossing the Lower Bridge.
3. The 2nd and 20th Georgia continue their withdrawal after their retreat from the bridge.
4. Brown and Reilly duel with Battery E, 4th United States and Battery D, Pennsylvania Light Artillery.

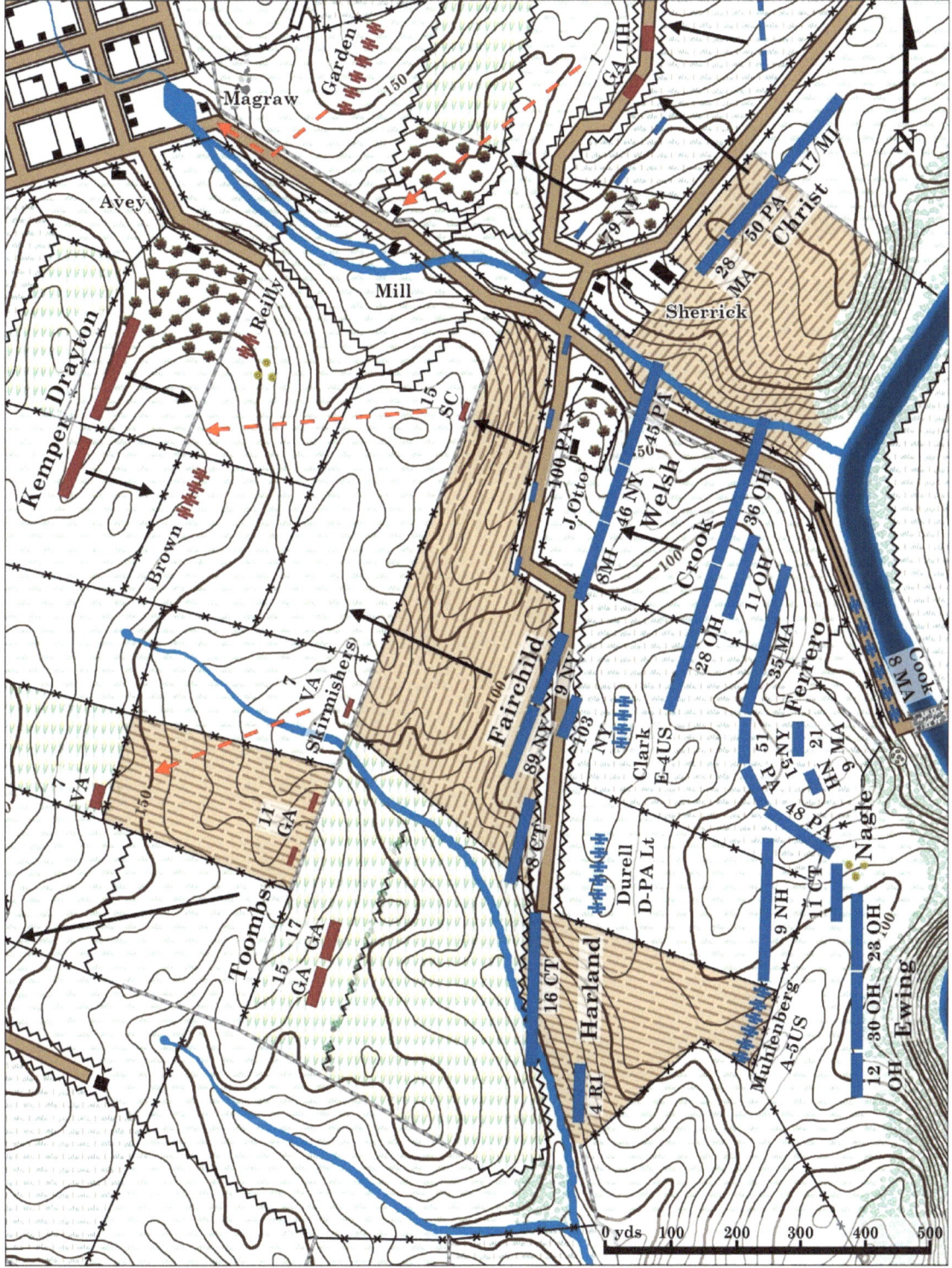

1. 3:30 to 3:45 p.m.
2. Willcox's division advances, both brigades pushing their opposing Confederate skirmishers away.
3. Fairchild advances to take and silence Brown and Reilly, pushing the 7th Virginia skirmishers back.[1]
4. Kemper and Drayton, hidden in a ravine and not visible to the Federals, advance to support Brown and Reilly.[2]
5. The 15th and 17th Georgia fall back to regroup with the rest of the brigade.

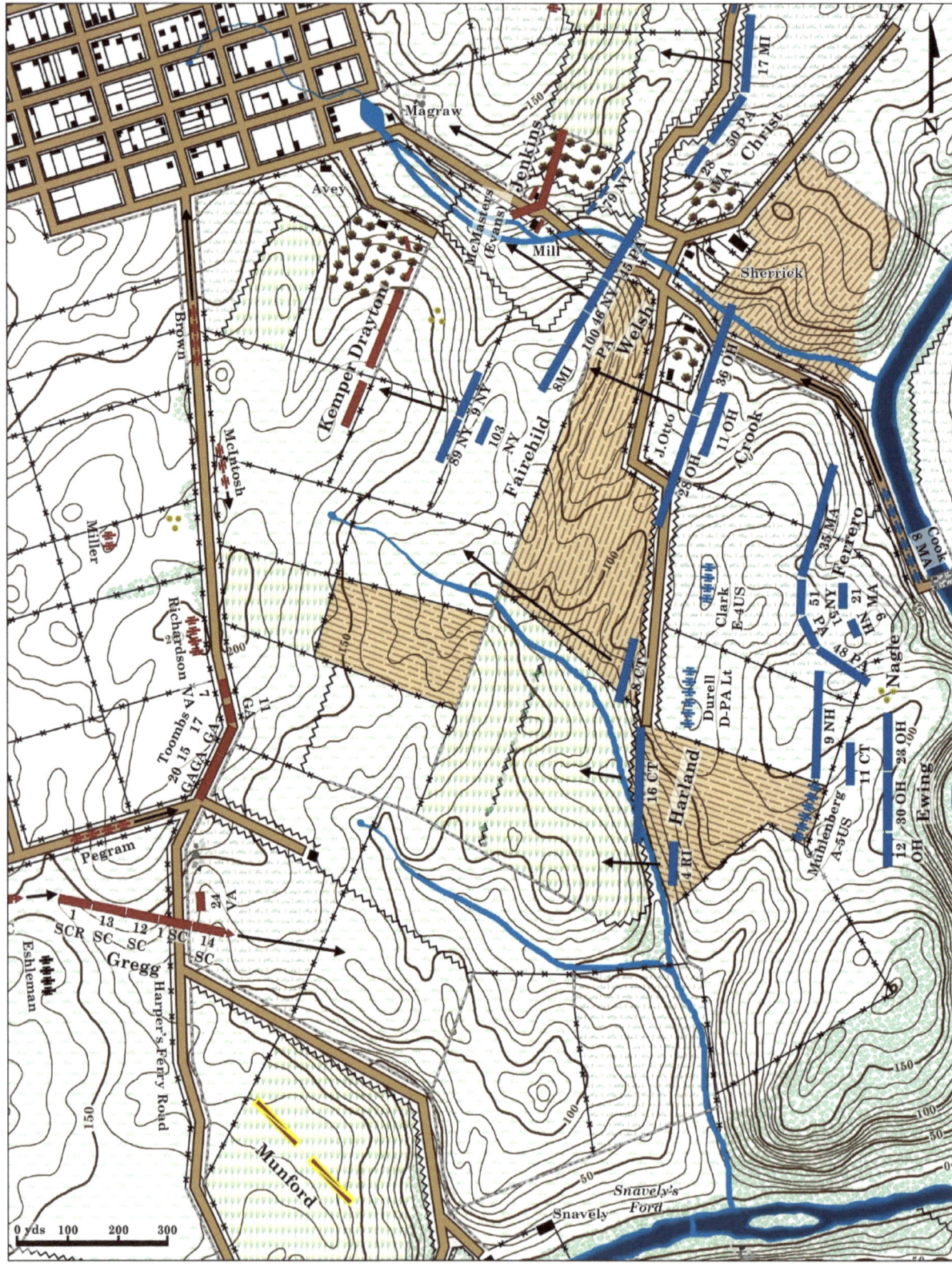

1. 3:50 p.m.
2. Willcox and Rodman's divisions continue their advance.
3. A. P. Hill's Light Division begins to arrive, led by Gregg's Brigade.

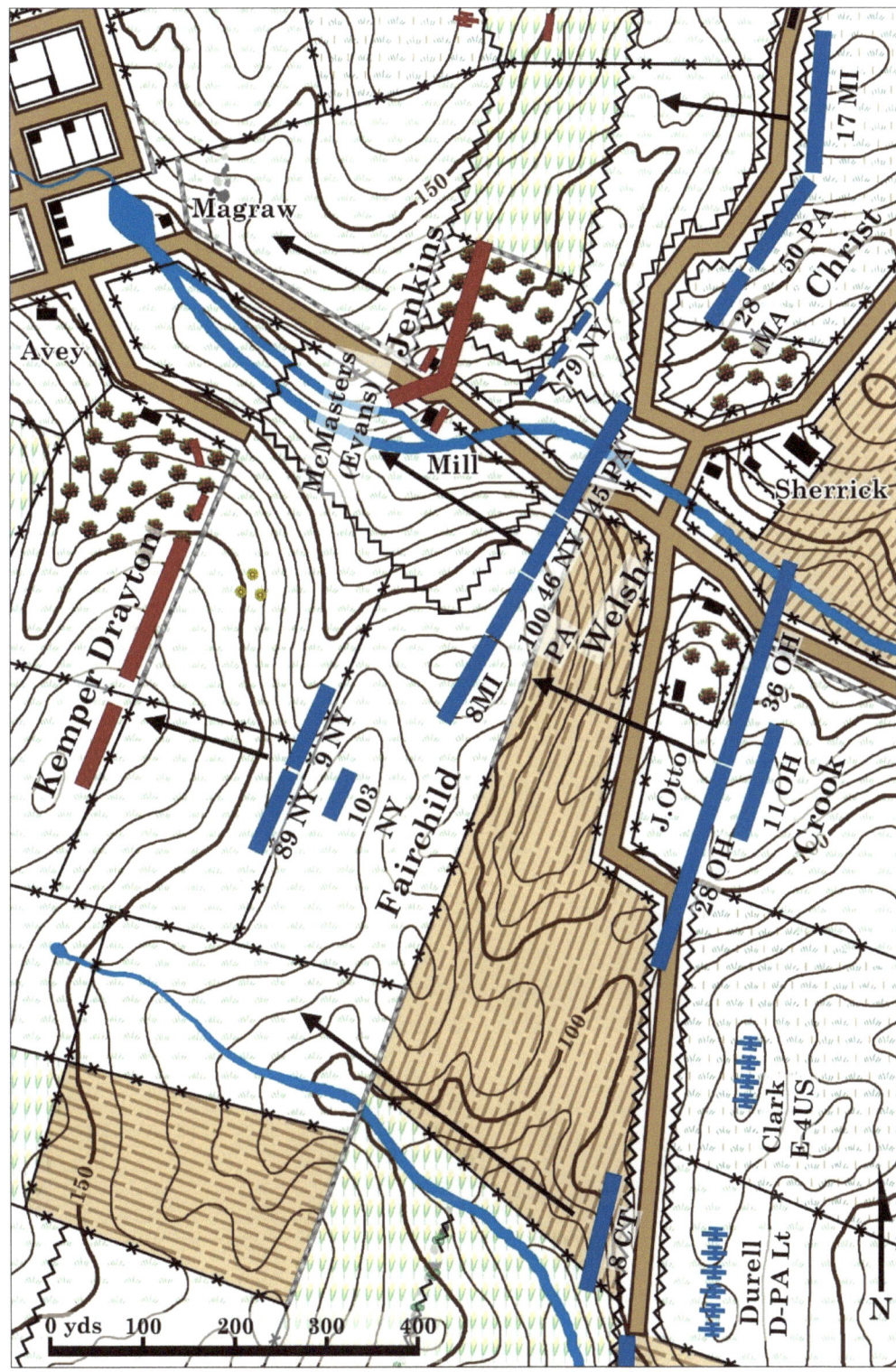

1. 3:50 p.m.
2. The 17th Michigan moves forward.³
3. Welsh's advance drives back the skirmishers at the mill and forces Jenkin's Brigade to withdraw.⁴
4. Brown and Reilly withdraw from Fairchild's advance. Fairchild charges the wall against Dayton and Kemper.

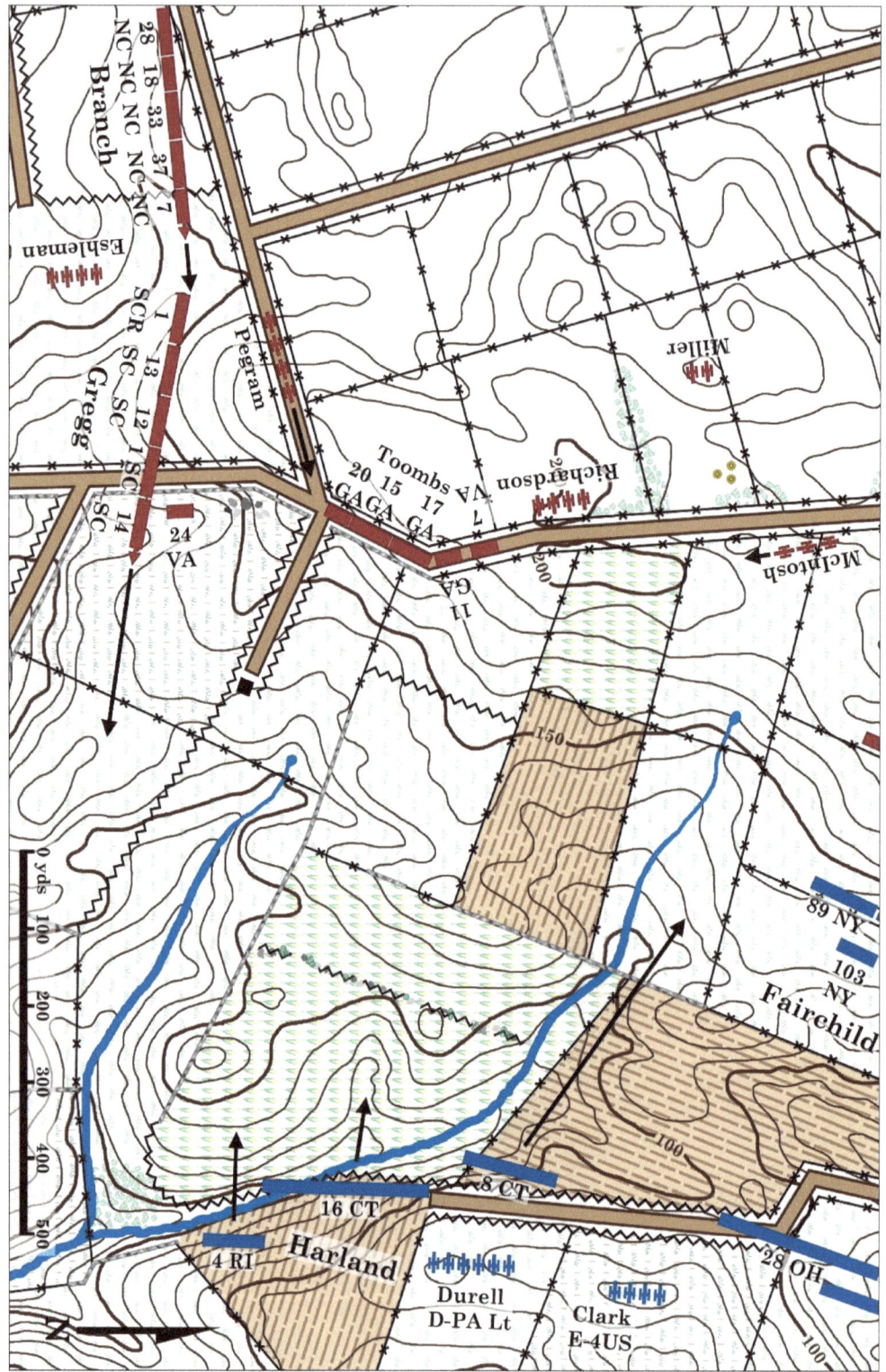

1. 3:50 p.m.
2. Harland's brigade advances. The 8th Connecticut drifts to the right. The 16th Connecticut and 4th Rhode Island hesitate because they can see Confederates arriving on their left flank.[5]
3. Gregg's Brigade, the lead infantry in A. P. Hill's Light Division, arrives and crosses the Harper's Ferry Road.

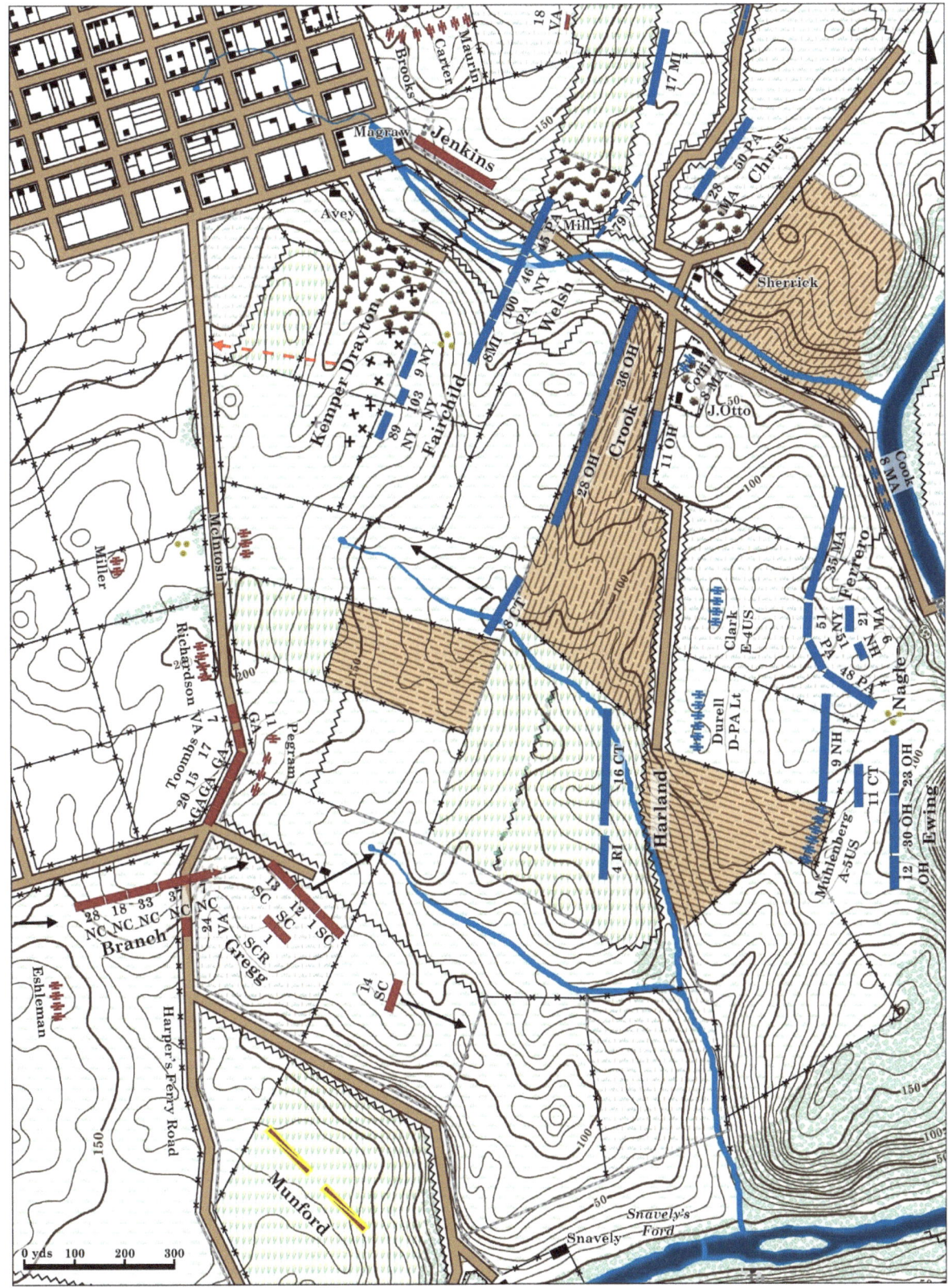

1. 4:00 p.m.
2. Willcox and Fairchild reach the limit of their advance.
3. A. P. Hill's Division continues to deploy.

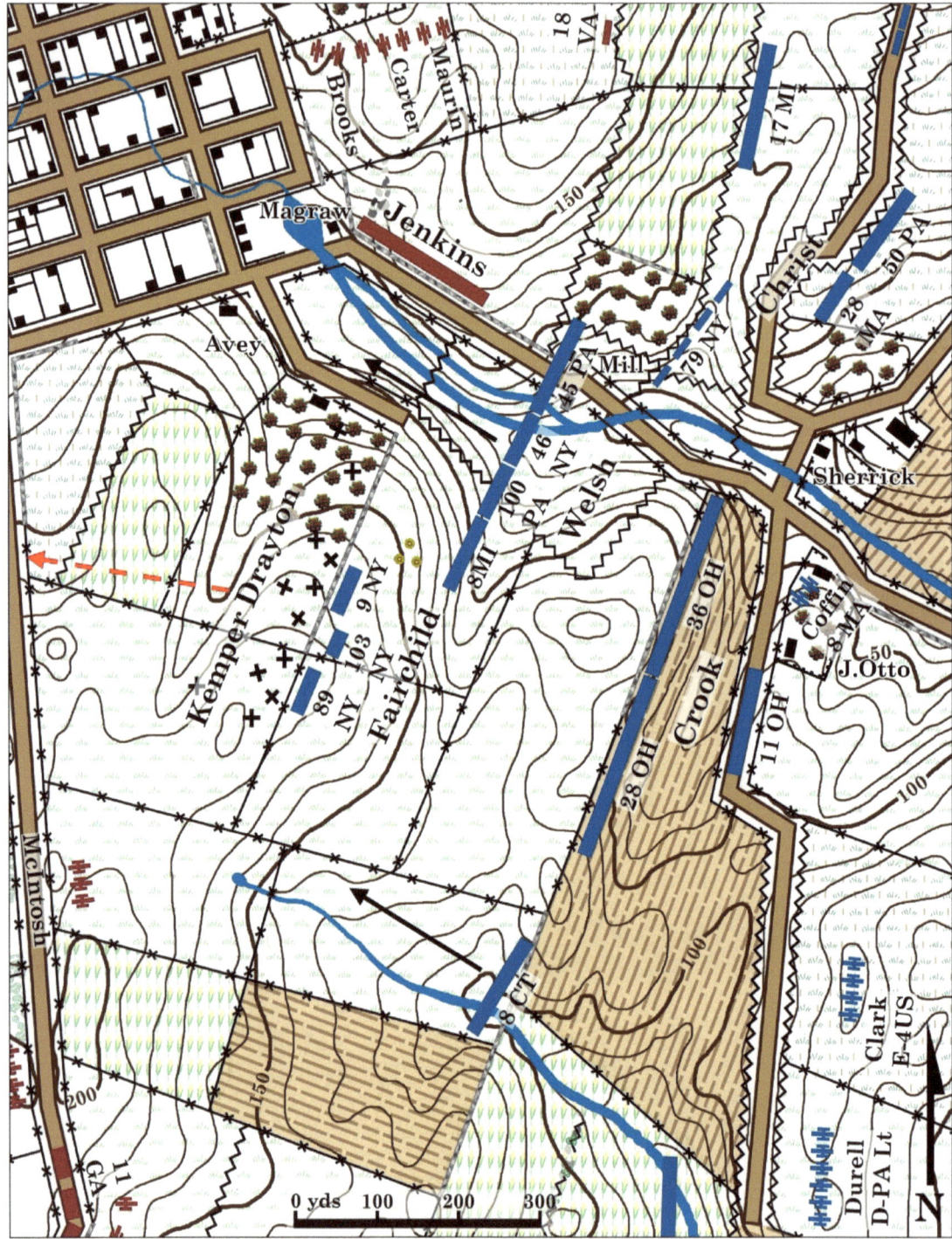

1. 4:00 p.m.
2. Welsh continues to advance up the creek past the mill. The 45th Pennsylvania remains at the mill, but the left wing of the brigade continues further into the orchard.[6]
3. Fairchild charges the stone wall and routs Kemper and Drayton, who fall back to the Harper's Ferry Road.[7]
4. The 8th Connecticut continues to advance without the rest of Harland's brigade. Harland and Rodman are with the regiment.[8]
5. Jenkin's Brigade is protected by Christ and the 45th Pennsylvania by the reverse slope of the hill.

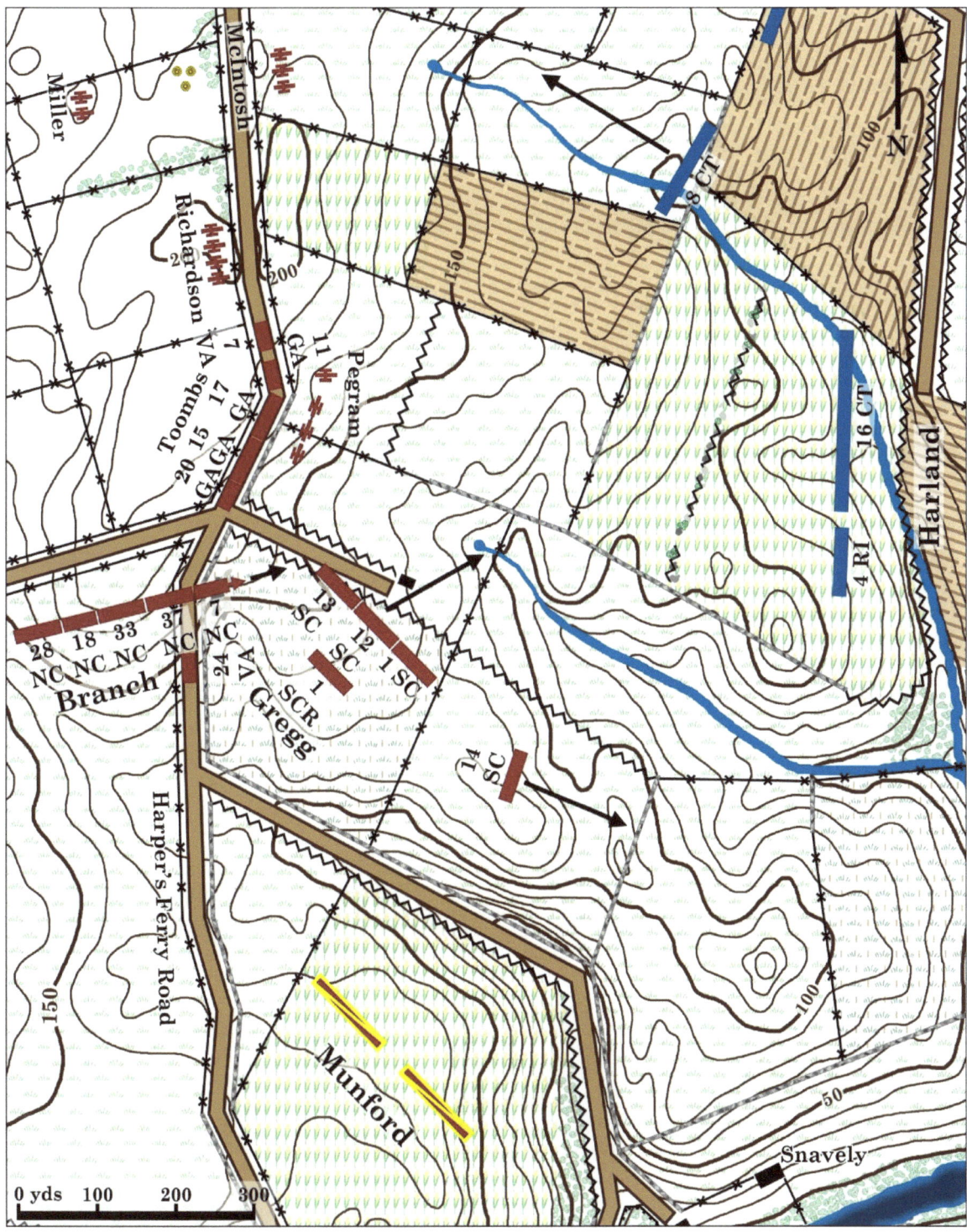

1. 4:00 p.m.
2. The 8th Connecticut continues forward, but the rest of the brigade halts to confront Gregg.
3. Gregg's Brigade deploys into line and moves toward the southwest corner of the Otto 40 acre cornfield.[9]
4. The 14th South Carolina continues east to protect the brigade flank.[10]
5. Branch's Brigade moves up to the farm lane east of the Harper's Ferry Road behind Gregg.

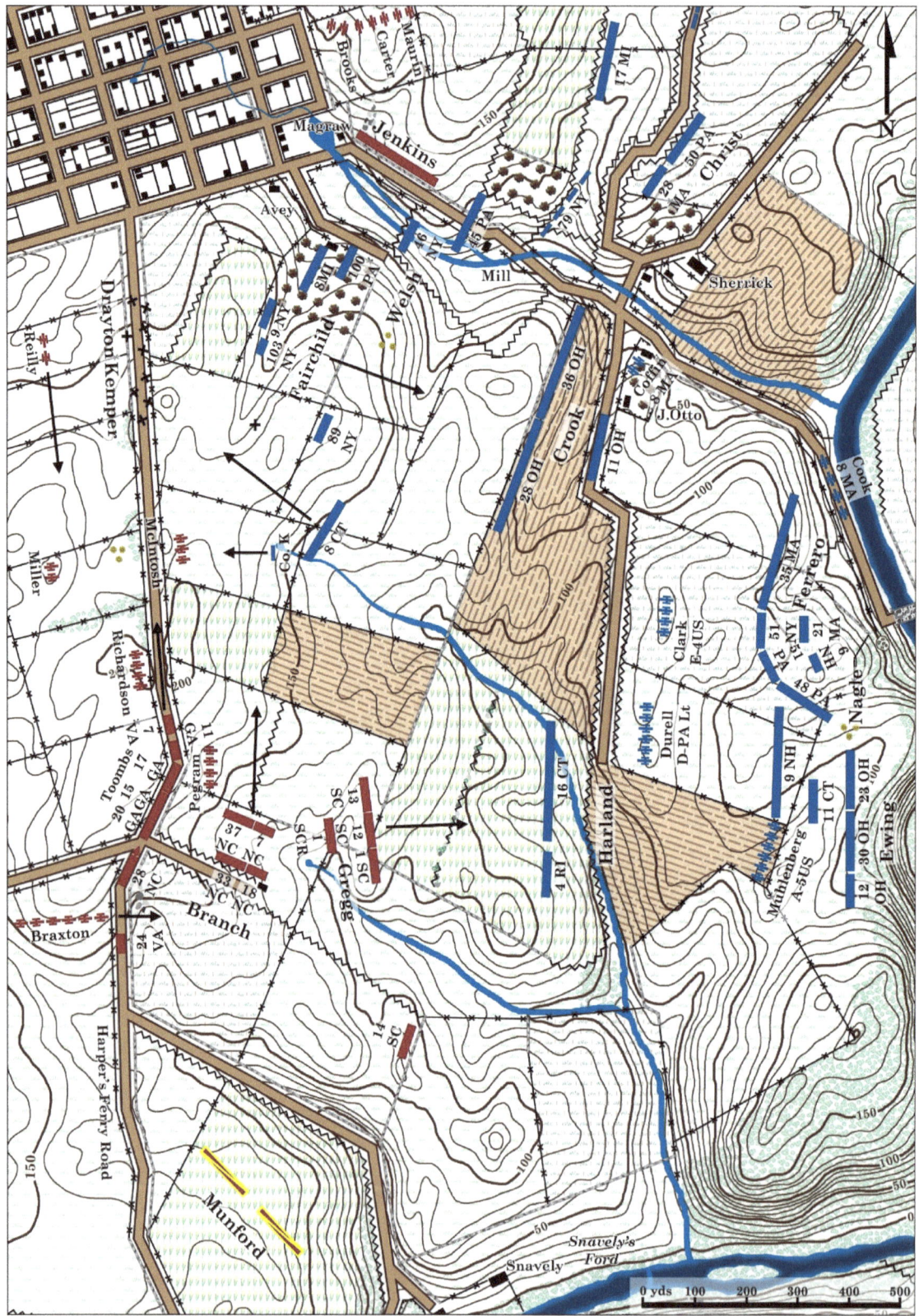

1. 4:15 p.m.
2. Willcox and Fairchild reach the apex of their advance.
3. A. P. Hill's Light Division continues to arrive and deploy.

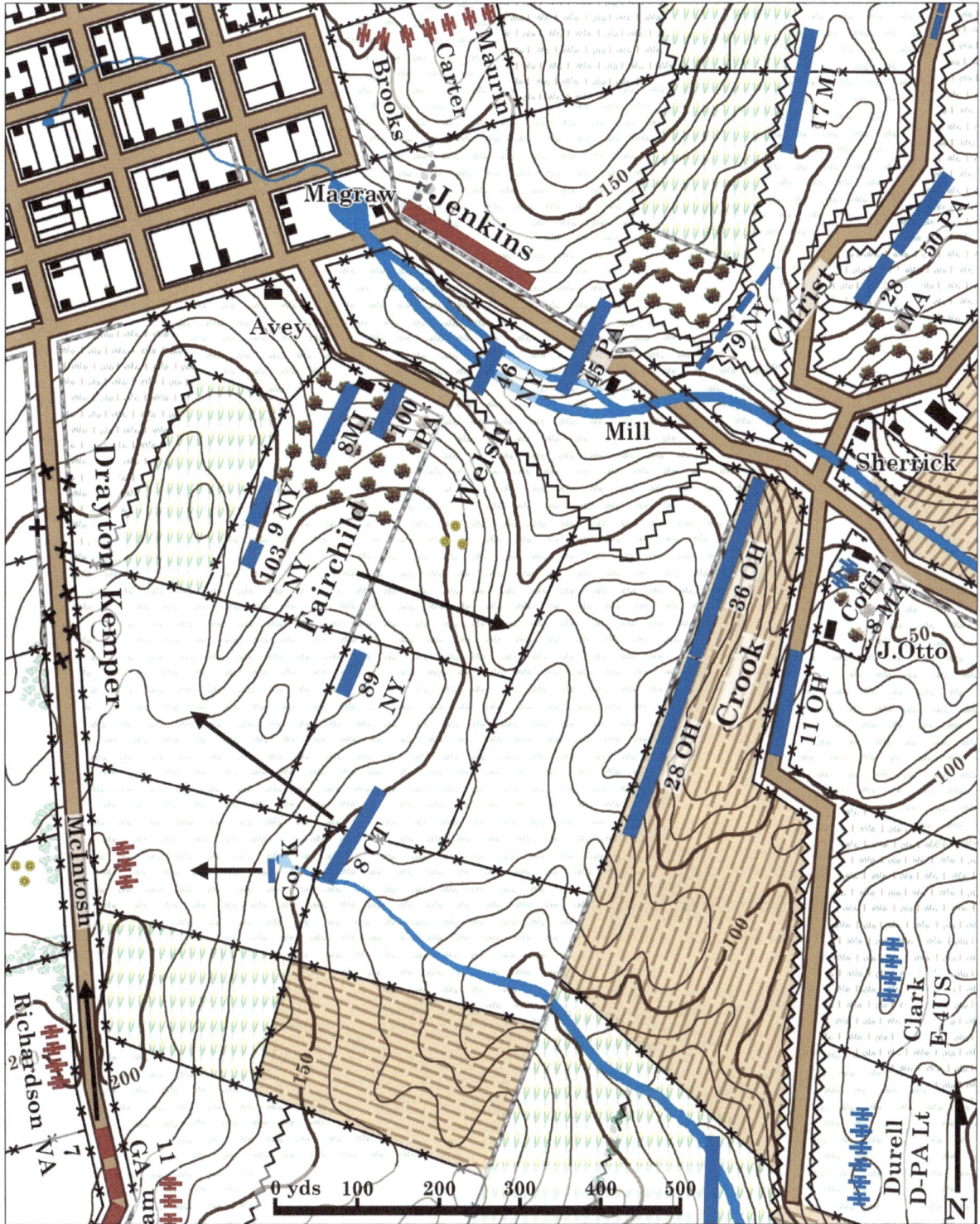

1. 4:15 p.m.
2. Some of Fairchild and Welsh's men reach Sharpsburg. One member of Welsh's brigade is reported to have been killed in the street north of the Avey house.[11]
3. Fairchild falls back after a short while.
4. The 8th Connecticut continues forward. Company K is detached to charge McIntosh's Battery. The gunners abandon the cannon and retreat.[12]

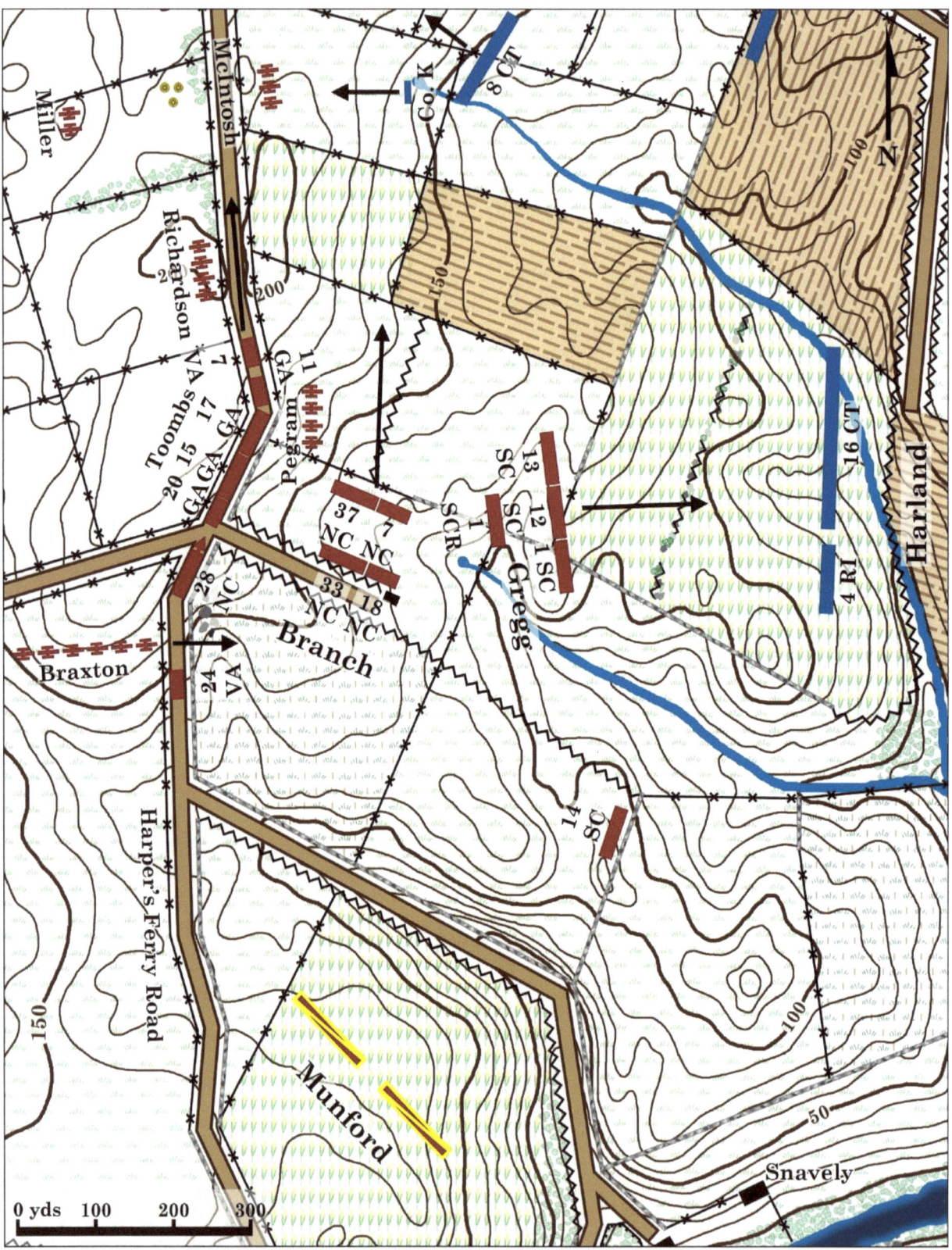

1. 4:15 p.m.
2. Gregg's Brigade moves into the Otto cornfield.
3. The 7th and 37th North Carolina of Branch's Brigade move north.

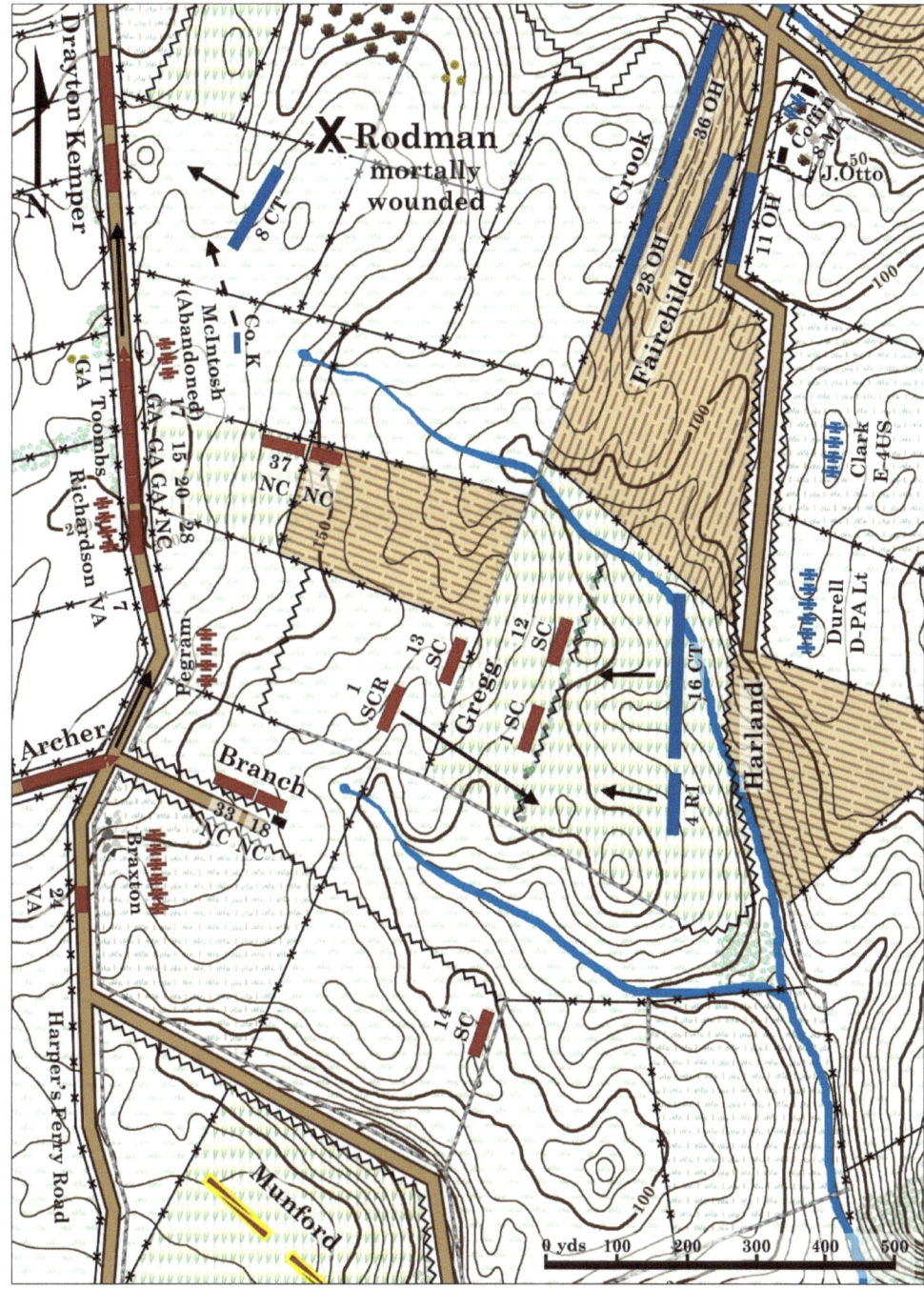

1. 4:20 p.m.
2. The 8th Connecticut moves to engage Kemper and Drayton.
3. Company K moves back to rejoin the regiment.[13]
4. The 7th and 37th North Carolina engage the 8th Connecticut on the flank.[14]
5. Toombs' Brigade moves north on the Harper's Ferry Road to engage the 8th Connecticut.[15]
6. General Rodman is mortally wounded riding back to the balance of Harland's brigade in an effort to move them forward.[16]
7. The 1st and 12th South Carolina fire into Harland at the bottom of the ravine. The 16th Connecticut and 4th Rhode Island move forward and push them back.[17]
8. The 1st South Carolina Rifles marches past the end of Harland's line.[18]

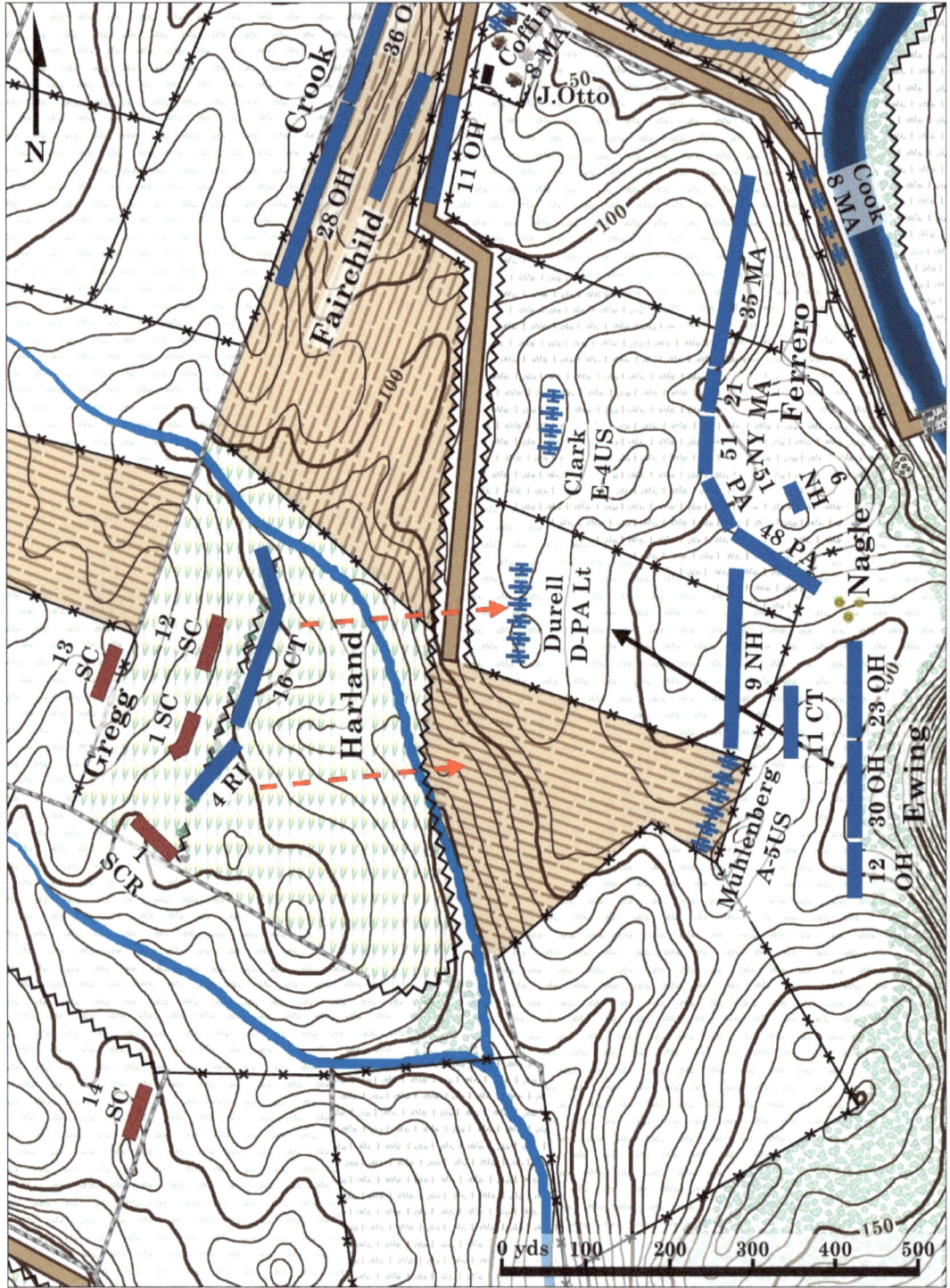

1. 4:25 p.m.
2. Gregg's South Carolinians catch the 16th Connecticut and 4th Rhode Island in front and flank. The 4th Rhode Island flanks the 1st South Carolina, but are in turn flanked by the 1st South Carolina Rifles.[19]
3. The 16th attempts maneuvers too complicated for the green regiment and both units rout out of the cornfield.[20]
4. Ewing's brigade moves forward.

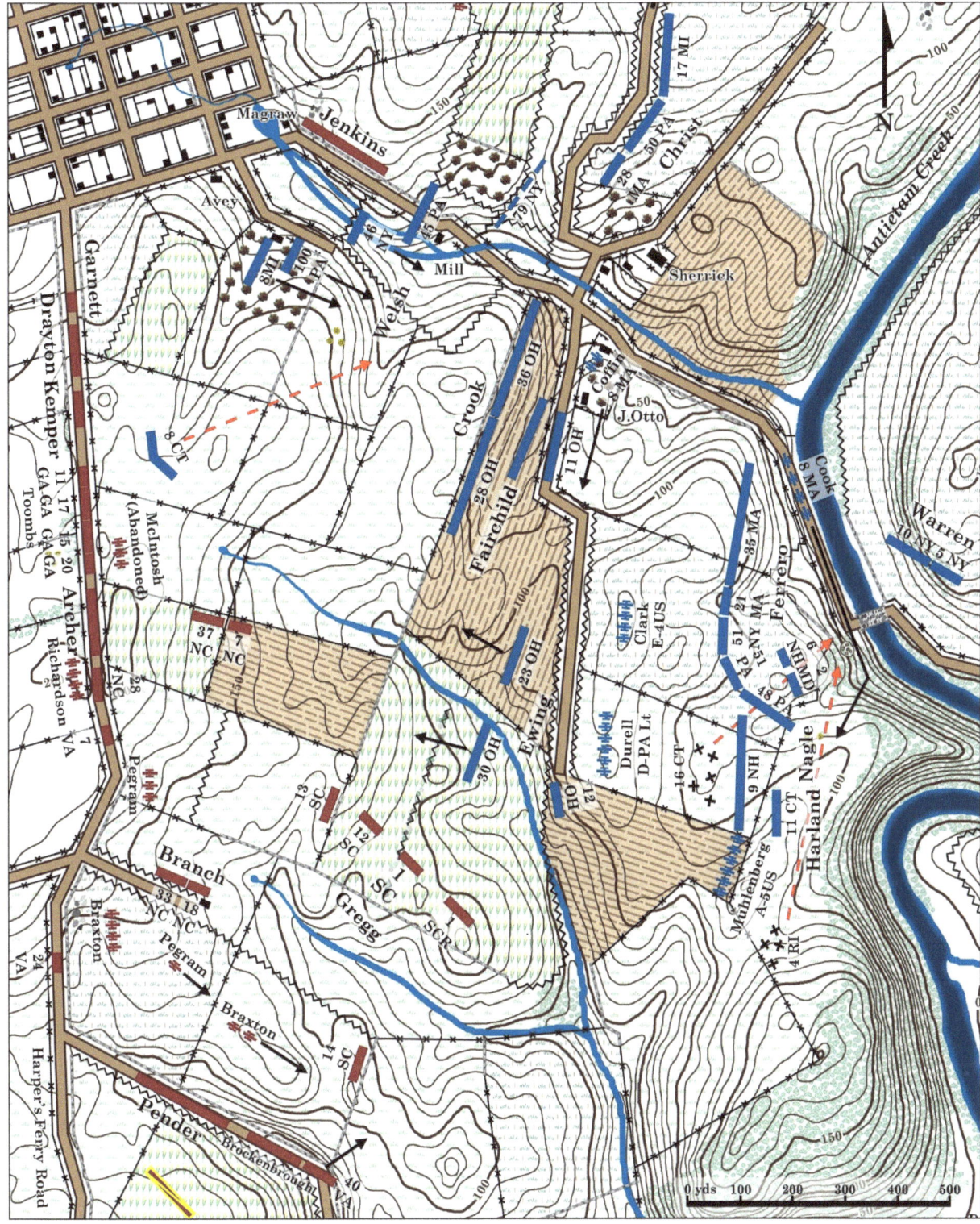

1. 4:30 p.m.
2. Welsh's brigade falls back from the Avey orchard.
3. Exposed, alone, and taking fire from three directions, the 8th Connecticut retreats.[21]
4. Ewing's brigade crosses the Otto cornfield toward the stone wall.[22]
5. Pender and Brockenbrough secure the Confederate army's right flank.

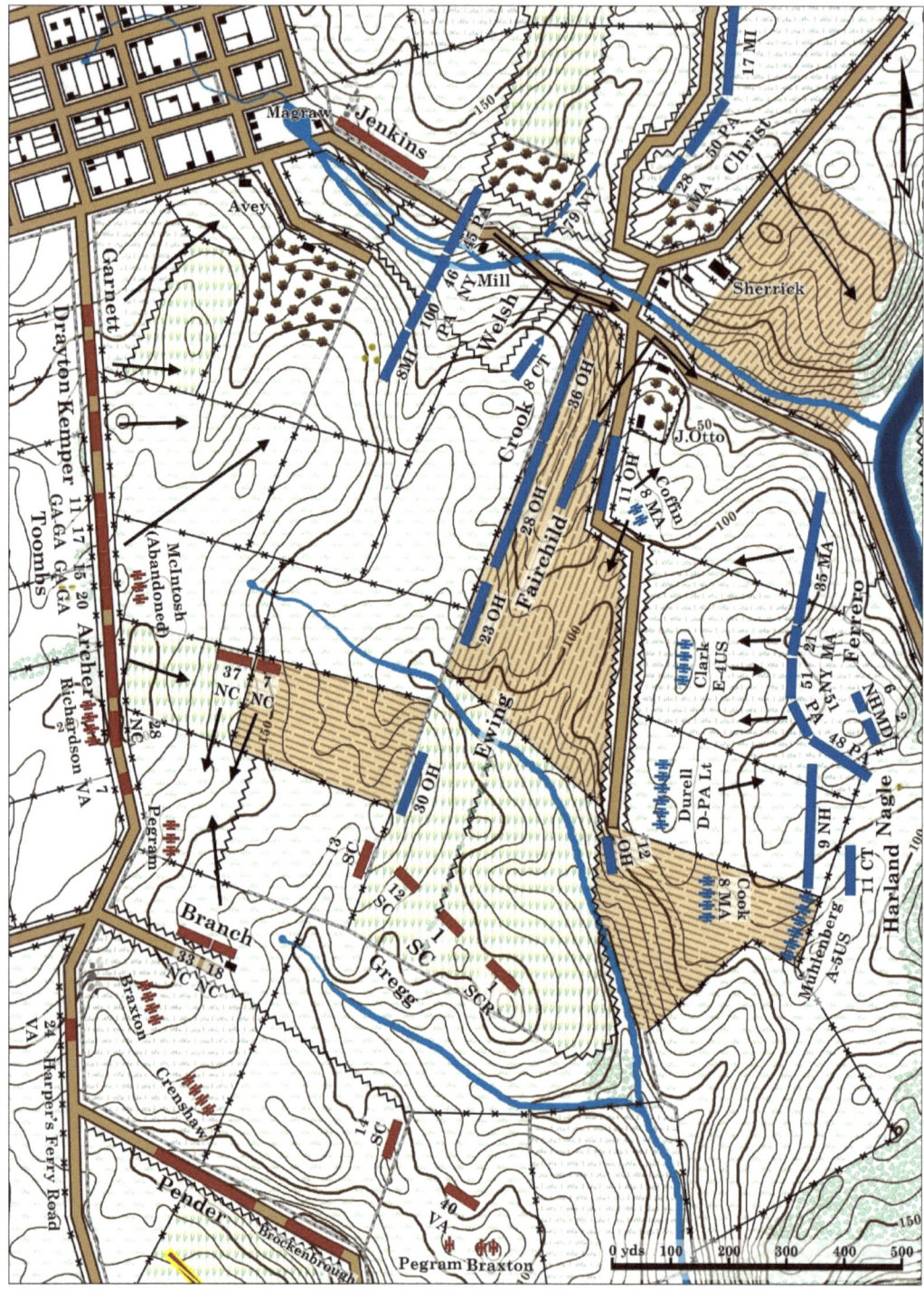

1. 4:40 p.m.
2. Willcox's division falls back toward the creek.
3. Batteries E, 4th United States and D, Pennsylvania Light retire from the ridge and are replaced by regiments from Sturgis division.[23]
4. Ewing's brigade takes a position at the stone wall. Because of the high corn, they don't see Gregg's Confederates nearby.
5. Branch's Brigade moves to reunite.
6. The Confederate line in the Harper's Ferry Road advances toward the creek and bridge.

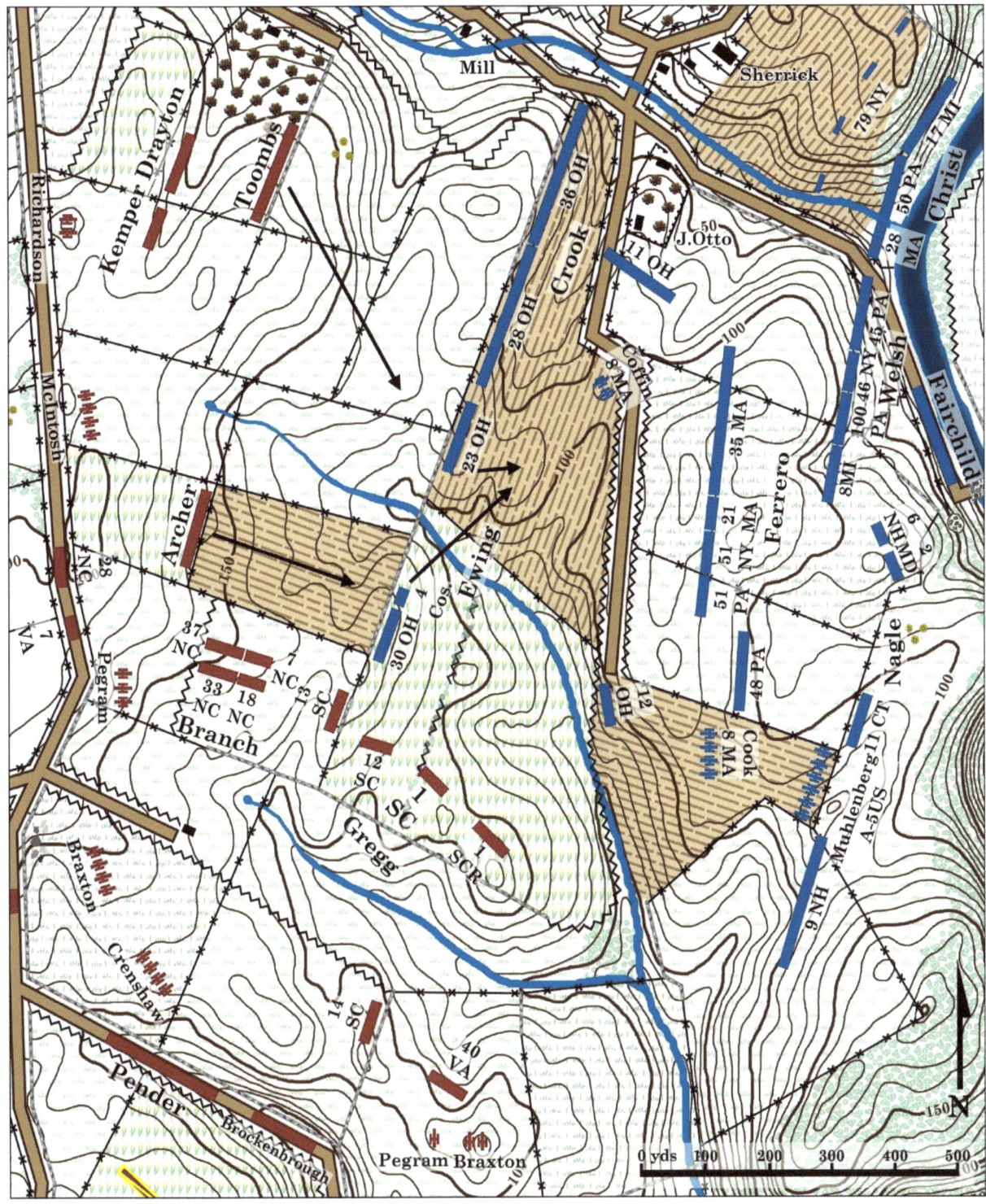

1. 4:50 p.m.
2. Archer and Toombs' Brigades continue advancing to the stone wall.[24]
3. Ewing orders his brigade to fall back. Only four companies of the 30th Ohio hear and obey the order, leaving the remainder of the regiment at the stone wall.[25]

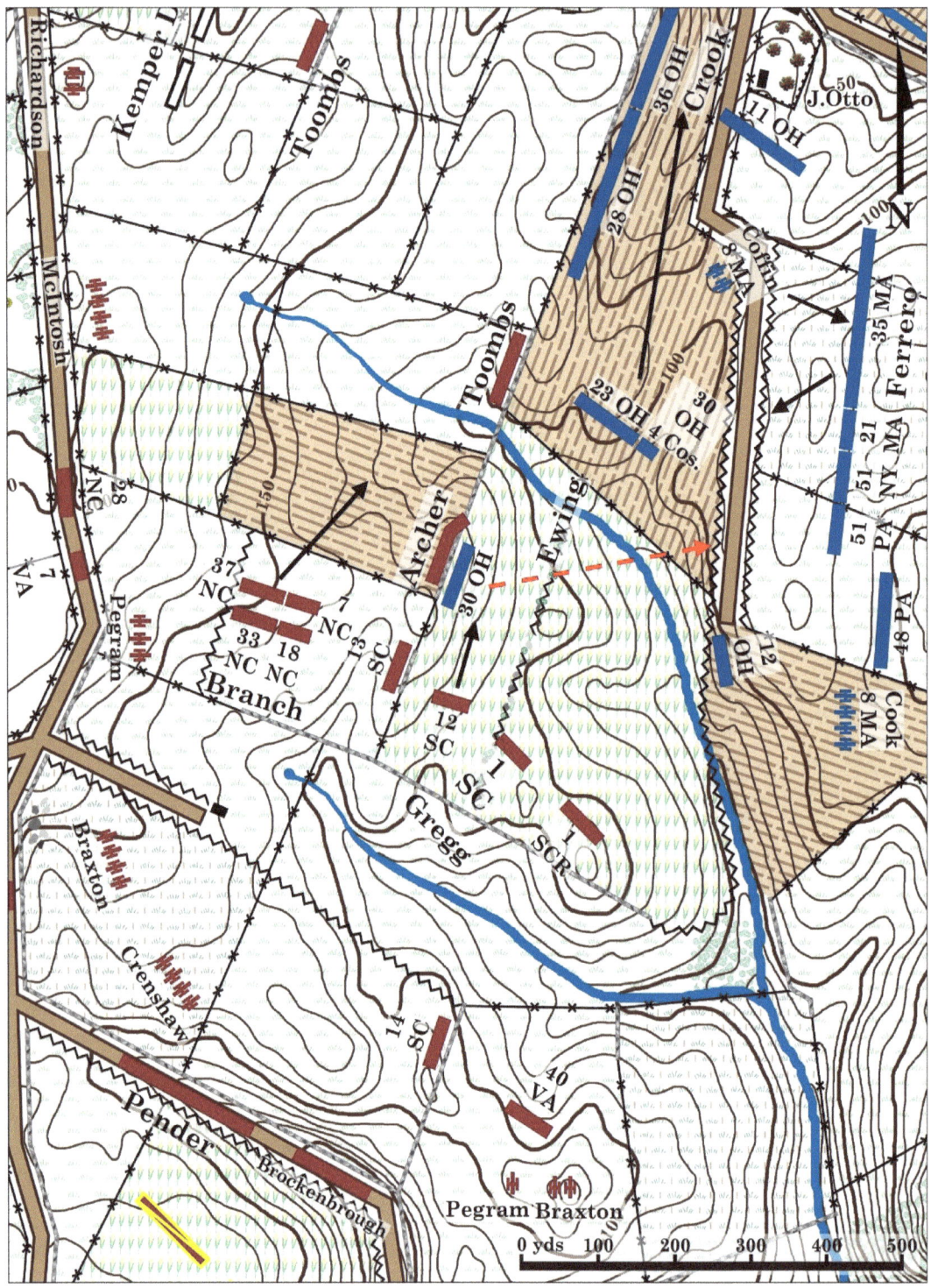

1. 5:00 p.m.
2. Archer confronts the 30th Ohio, even advancing over the stone wall on their right.[26]
3. The 12th South Carolina charges the 30th Ohio's flank. The Ohioans break and flee to the rear.[27]
4. The 23rd Ohio and the four companies of the 30th Ohio fall back to the north.[28]
5. The 35th Massachusetts advances to the Otto farm lane. The rest of Ferrero's brigade lays down on the ridge and engages the Confederates at long range.[29]
6. Branch's Brigade moves north.

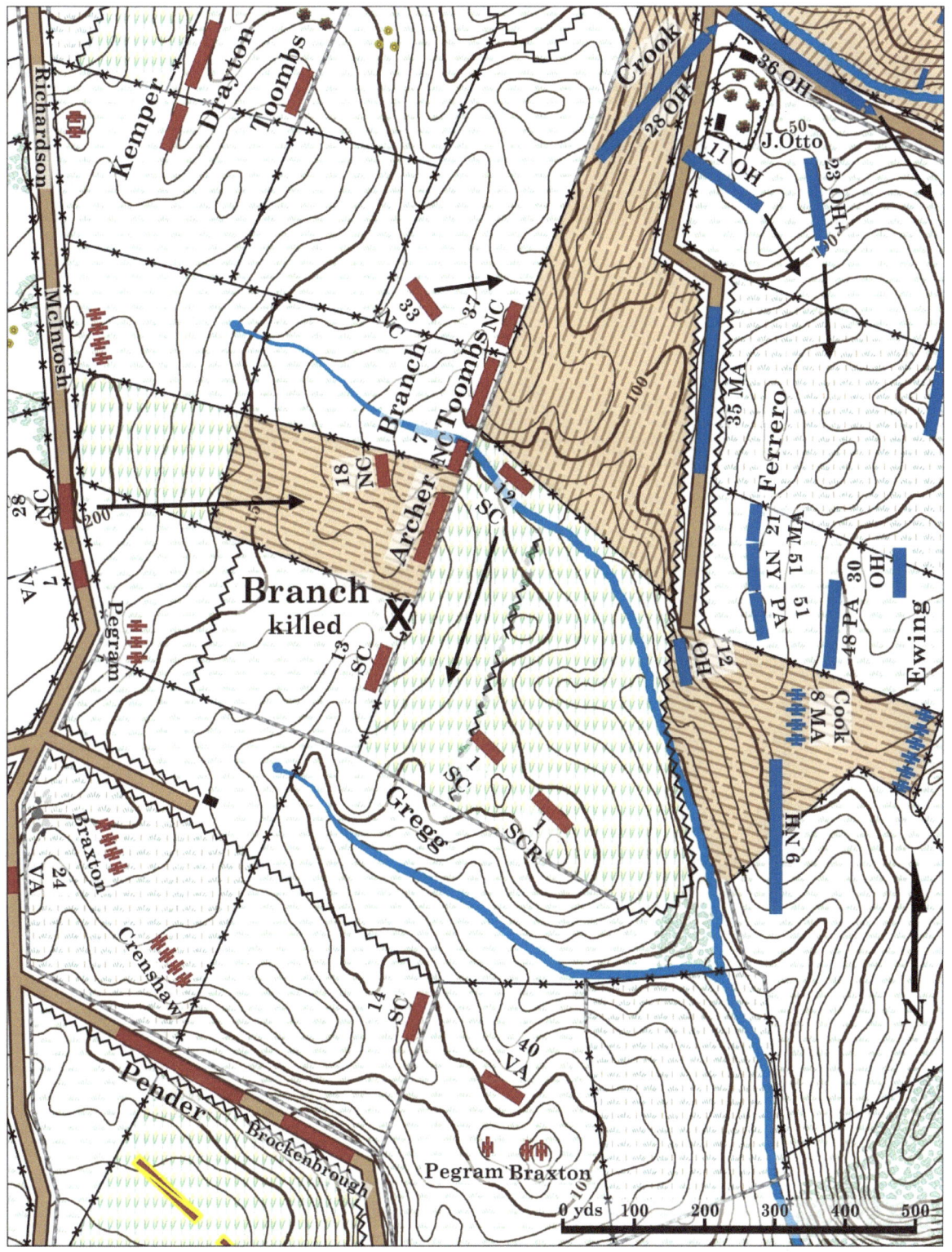

1. 5:15 p.m.
2. Branch's Brigade fills in gaps at the stone wall.
3. General Branch is killed while observing the Union lines.[30]
4. The 12th South Carolina falls back.
5. The 35th Massachusetts and Confederates at the wall engage each other until nightfall.[31]

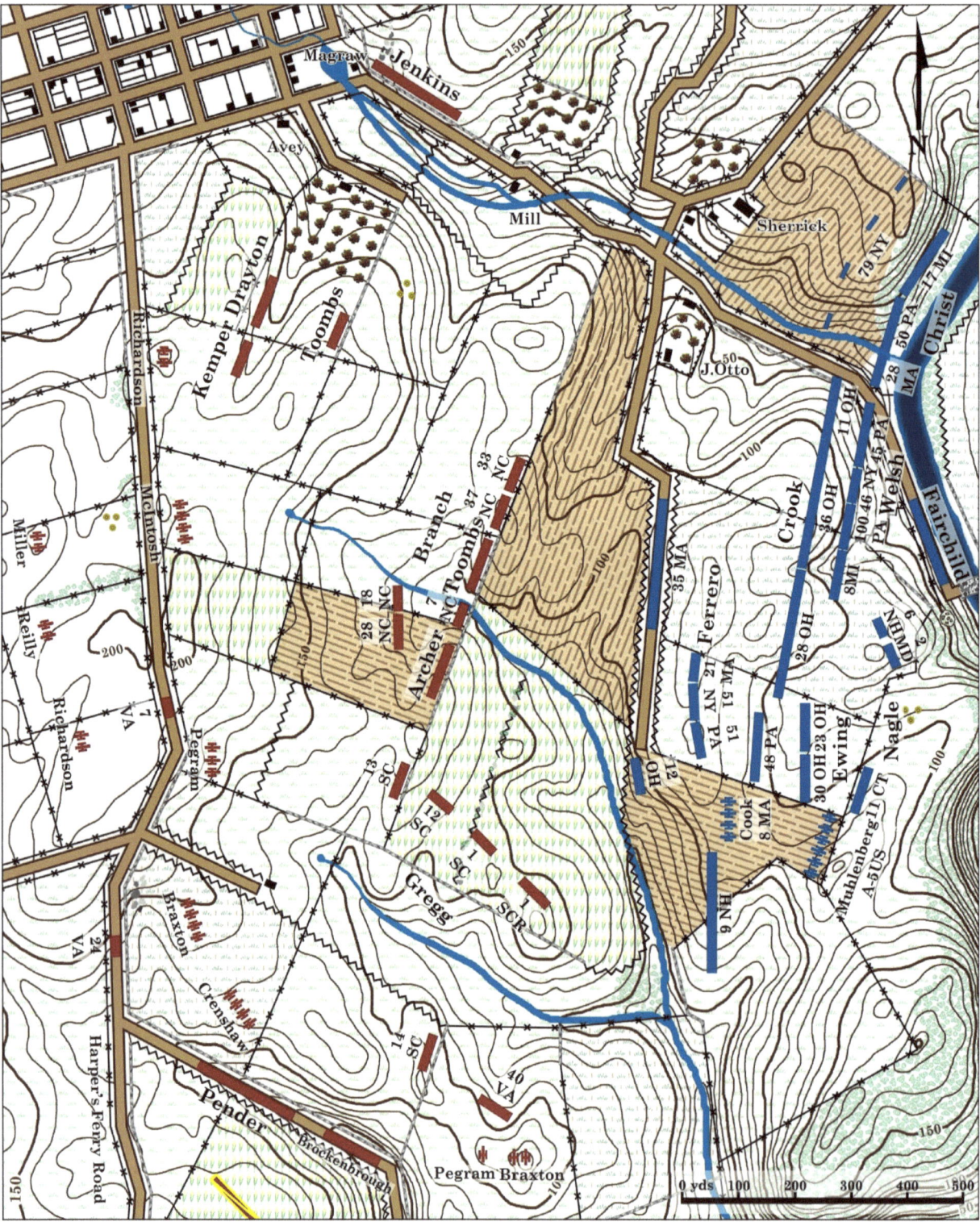

1. 5:30 p.m.
2. The opposing lines below Sharpsburg near nightfall.
3. At or after nightfall the Ninth Corps would re-arrange its lines for the night.

Order of Battle

Union

Army of the Potomac
Major General George B. McClellan

General Headquarters

Escort
Independent Company Oneida (New York) Cavalry
4th United States Cavalry, Companies A and E

Provost Guard
Major William F. Hood
2nd United States Cavalry
8th United States
19th United States

Headquarters Guard
93rd New York

First Army Corps
Major General Joseph Hooker

1st Division
Brigadier General Abner Doubleday

1st Brigade	2nd Brigade	3rd Brigade
Colonel Walter Phelps Jr.	Lt. Colonel J. William Hofmann	Brigadier General Marsena R. Patrick
22nd New York	7th Indiana	21st New York
24th New York	76th New York	23rd New York
30th New York	95th New York	35th New York
84th New York*	56th Pennsylvania	80th New York
2nd United States Sharpshooters		
*Also known as the 14th Brooklyn		

4th Brigade	Artillery
Brigadier General John Gibbon	1st New Hampshire Battery
19th Indiana	Battery D, 1st Rhode Island
2nd Wisconsin	Battery L, 1st New York
6th Wisconsin	Battery B, 4th United States
7th Wisconsin	

2d Division
Brigadier General James B. Ricketts

1st Brigade	2nd Brigade	3rd Brigade
Brigadier General Abram Duryée	Colonel William A. Christian	Brigadier General George L. Hartsuff
97th New York	26th New York	12th Massachusetts
104th New York	94th New York	13th Massachusetts
105th New York	88th Pennsylvania	83rd New York
107th Pennsylvania	90th Pennsylvania	11th Pennsylvania

Artillery
Battery F, 1st Pennsylvania
Battery C, Pennsylvania Light

3rd Division
Brigadier General George G. Meade

1st Brigade	2nd Brigade	3rd Brigade
Brigadier General Truman Seymour	Colonel Albert L. Magilton	Lt. Colonel Robert Anderson
1st Pennsylvania Reserves	3rd Pennsylvania Reserves	9th Pennsylvania Reserves
2nd Pennsylvania Reserves	4th Pennsylvania Reserves	10th Pennsylvania Reserves
5th Pennsylvania Reserves	7th Pennsylvania Reserves	11th Pennsylvania Reserves
6th Pennsylvania Reserves	8th Pennsylvania Reserves	12th Pennsylvania Reserves
13th Pennsylvania Reserves		

Artillery
Battery A, 1st Pennsylvania
Battery B, 1st Pennsylvania
Battery C, 5th United States

Second Army Corps
Major General Edwin V. Sumner

1st Division
Major General Israel B. Richardson

1st Brigade	2nd Brigade	3rd Brigade
Brigadier General John C. Caldwell	Brigadier General Thomas F. Meagher	Colonel John R. Brooke
5th New Hampshire	29th Massachusetts	2nd Delaware
7th New York	63rd New York	52nd New York
61st & 64th New York	69th New York	57th New York
81st Pennsylvania	88th New York	66th New York
		53rd Pennsylvania

Artillery
Battery B, 1st New York
Batteries A&C, 4th United States

2nd Division
Major General John Sedgwick

<u>1st Brigade</u>
Brigadier General Willis A. Gorman
15th Massachusetts
1st Minnesota
34th New York
82nd New York
Massachusetts Sharpshooters, 1st Company
Minnesota Sharpshooters, 2nd Company

<u>Artillery</u>
Battery A, 1st Rhode Island
Battery I, 1st United States

<u>2nd Brigade</u>
Brigadier General Oliver O. Howard
69th Pennsylvania
71st Pennsylvania
72nd Pennsylvania
106th Pennsylvania

<u>3rd Brigade</u>
Brigadier General Napoleon J. T. Dana
19th Massachusetts
20th Massachusetts
7th Michigan
42nd New York
59th New York

3rd Division
Brigadier General William H. French

<u>1st Brigade</u>
Brigadier General Nathan Kimball
14th Indiana
8th Ohio
132nd Pennsylvania
7th West Virginia

<u>Unattached Artillery</u>
Battery G, 1st New York
Battery B, 1st Rhode Island
Battery G, 1st Rhode Island

<u>2nd Brigade</u>
Colonel Dwight Morris
14th Connecticut
108th New York
130th Pennsylvania

<u>3rd Brigade</u>
Brigadier General Max Weber
1st Delaware
5th Maryland
4th New York

Fifth Army Corps
Major General Fitz John Porter

1st Division
Major General George W. Morell

<u>1st Brigade</u>
Colonel James Barnes
2nd Maine
18th Massachusetts
22nd Massachusetts
1st Michigan
13th New York
25th New York
118th Pennsylvania
Massachusetts Sharpshooters, 2nd Company

<u>2nd Brigade</u>
Brigadier General Charles Griffin
2nd District of Columbia
9th Massachusetts
32nd Massachusetts
4th Michigan
14th New York
62nd Pennsylvania

<u>3rd Brigade</u>
Colonel Thomas B. W. Stockton
20th Maine
16th Michigan
12th New York
17th New York
44th New York
83rd Pennsylvania
Michigan Sharpshooters, Brady's Company

Unassigned
1st United States Sharpshooters

Artillery
Battery C, Massachusetts Light
Battery C, 1st Rhode Island
Battery D, 5th United States

2nd Division
Brigadier General George Sykes

1st Brigade
Lt. Colonel Robert C. Buchanan
3rd United States
4th United States
12th United States, 1st Battalion
12th United States, 2nd Battalion
14th United States, 1st Battalion
14th United States, 2nd Battalion

2nd Brigade
Major Charles S. Lovell
1st & 6th United States
2nd & 10th United States
11th United States
17th United States

3rd Brigade
Colonel Gouverneur K. Warren
5th New York
10th New York

Artillery
Batteries E & G, 1st United States
Battery I, 5th United States
Battery K, 5th United States

Artillery Reserve
Battery A, 1st Battalion New York
Battery B, 1st Battalion New York
Battery C, 1st Battalion New York
Battery D, 1st Battalion New York
5th New York Battery

Battery K, 1st United States
Battery G, 4th United States

Sixth Army Corps
Major General William B. Franklin

1st Division
Major General Henry W. Slocum

1st Brigade
Colonel Alfred T. A. Torbert
1st New Jersey
2nd New Jersey
3rd New Jersey
4th New Jersey

2nd Brigade
Colonel Joseph J. Bartlett
5th Maine
16th New York
27th New York
96th Pennsylvania

3rd Brigade
Brigadier General John Newton
18th New York
31st New York
32nd New York
95th Pennsylvania

Artillery
Battery A, Maryland Light
Battery A, Massachusetts Light
Battery A, New Jersey Light
Battery D, 2nd United States

2nd Division
Major General William F. Smith

1st Brigade	2nd Brigade	3rd Brigade
Brigadier General Winfield S. Hancock	Brigadier General William T. H. Brooks	Colonel William H. Irwin
6th Maine	2nd Vermont	7th Maine
43rd New York	3rd Vermont	20th New York
49th Pennsylvania	4th Vermont	33rd New York
137th Pennsylvania	5th Vermont	49th New York
5th Wisconsin	6th Vermont	77th New York

Artillery
Battery B, Maryland Light
1st New York Light Battery
Battery F, 5th United States

Ninth Army Corps
Major General Ambrose Burnside
Brigadier General Jacob D. Cox

1st Division
Brigadier General Orlando B. Willcox

1st Brigade	2nd Brigade	Artillery
Colonel Benjamin C. Christ	Colonel Thomas Welsh	8th Battery Massachusetts Light
28th Massachusetts	8th Michigan	Battery E, 2nd United States
17th Michigan	46th New York	
79th New York	45th Pennsylvania	
50th Pennsylvania	100th Pennsylvania	

2nd Division
Brigadier General Samuel D. Sturgis

1st Brigade	2nd Brigade	Artillery
Brigadier General James Nagle	Brigadier General Edward Ferrero	Battery D, Pennsylvania Light
2nd Maryland	21st Massachusetts	Battery E, 4th United States
6th New Hampshire	35th Massachusetts	
9th New Hampshire	51st New York	
48th Pennsylvania	51st Pennsylvania	

3rd Division
Brigadier General Isaac P. Rodman

1st Brigade	2nd Brigade	Artillery
Colonel Harrison S. Fairchild	Colonel Edward Harland	Battery A, 5th United States
9th New York	8th Connecticut	
89th New York	11th Connecticut	
103rd New York	16th Connecticut	
	4th Rhode Island	

Kanawha Division
Colonel Eliakim P. Scammon

1st Brigade
Colonel Hugh B. Ewing
12th Ohio
23rd Ohio
30th Ohio
Gilmore's Company West Virginia Cavalry
Harrison's Company West Virginia Cavalry
1st Battery Ohio Light Artillery

2nd Brigade
Colonel George Crook
11th Ohio
28th Ohio
36th Ohio

Unattached
6th New York Cavalry
Ohio Cavalry, 3d Independent Co.

Unattached Artillery
Batteries L & M 3rd United States
Battery L, 2nd New York

Twelfth Army Corps
Brigadier General Joseph K. Mansfield

1st Division
Brigadier General Alpheus S. Williams

1st Brigade
Brigadier General Samuel W. Crawford
10th Maine
28th New York
46th Pennsylvania
124th Pennsylvania
125th Pennsylvania
128th Pennsylvania

3rd Brigade
Brigadier General George H. Gordon
27th Indiana
2nd Massachusetts
13th New Jersey
107th New York
3rd Wisconsin

2nd Division
Brigadier General George S. Greene

1st Brigade	2nd Brigade	3rd Brigade
Lt. Colonel Hector Tyndale	Colonel Henry J. Stainrook	Colonel William B. Goodrich
5th Ohio	3rd Maryland	3rd Delaware
7th Ohio	102nd New York	Purnell Legion
66th Ohio	111th Pennsylvania	60th New York
28th Pennsylvania		78th New York

Artillery Battalion
4th Maine Battery
6th Maine Battery
Battery M, 1st New York
10th New York Battery
Battery E, Pennsylvania Light
Battery F, Pennsylvania Light
Battery F, 4th United States

Cavalry Division
Brigadier General Alfred Pleasonton

1st Brigade	2nd Brigade	3rd Brigade
Major Charles J. Whiting	Colonel John F. Farnsworth	Colonel Richard H. Rush
5th United States Cavalry	8th Illinois Cavalry	4th Pennsylvania Cavalry
6th United States Cavalry	3rd Indiana Cavalry	6th Pennsylvania Cavalry
	1st Massachusetts Cavalry	
	8th Pennsylvania Cavalry	

4th Brigade	5th Brigade	Artillery Battalion
Colonel Andrew T. McReynolds	Colonel Benjamin F. Davis	Battery A, 2nd United States
1st New York Cavalry	8th New York Cavalry	Batteries B & L, 2nd United States
12th Pennsylvania Cavalry	3rd Pennsylvania Cavalry	Battery M, 2nd United States
		Batteries C & G, 3rd United States

Confederate

Army of Northern Virginia
General Robert E. Lee

Longstreet's Command
Major General James Longstreet

McLaw's Division
Major General Lafayette McLaws

Kershaw's Brigade	Barksdale's Brigade	Semmes' Brigade
Brigadier General Joseph B. Kershaw	Brigadier General William Barksdale	Brigadier General Paul J. Semmes
2nd South Carolina	13th Mississippi	10th Georgia
3rd South Carolina	17th Mississippi	53rd Georgia
7th South Carolina	18th Mississippi	15th Virginia
8th South Carolina	21st Mississippi	32nd Virginia

Cobb's Brigade	Cabell's Battalion
Lt. Colonel Christopher C. Sanders	Manly's North Carolina Battery
16th Georgia	Pulaski Georgia Battery
24th Georgia	Richmond Fayette Artillery
Cobb's Legion	1st Richmond Howitzers
15th North Carolina	Troup Georgia Battery

Anderson's Division
Major General Richard H. Anderson

Wilcox's Brigade	Mahone's Brigade	Featherston's Brigade
Colonel Alfred Cumming	Colonel William A. Parham	Colonel Carnot Posey
8th Alabama	6th Virginia	12th Mississippi
9th Alabama	12th Virginia	16th Mississippi
10th Alabama	16th Virginia	19th Mississippi
11th Alabama	41st Virginia	2nd Mississippi Battalion

Armistead's Brigade	Pryor's Brigade	Wright's Brigade
Brigadier General Lewis A. Armistead	Brigadier General Roger A. Pryor	Brigadier General Ambrose R. Wright
9th Virginia	14th Alabama	44th Alabama
14th Virginia	2nd Florida	3rd Georgia
38th Virginia	8th Florida	22nd Georgia
53rd Virginia	3rd Virginia	48th Georgia
57th Virginia		

Saunders' Battalion
Donaldsonville Louisiana Battery
Norfolk Virginia Battery
Lynchburg Virginia Battery
Grimes's Virginia Battery

Jones' Division
Brigadier General David R. Jones

Garnett's Brigade
Brigadier General Richard B. Garnett
8th Virginia
18th Virginia
19th Virginia
28th Virginia
56th Virginia

Jenkins' Brigade
Colonel Joseph Walker
1st South Carolina
2nd South Carolina
5th South Carolina
6th South Carolina
4th South Carolina Bn.
Palmetto Sharpshooters

Kemper's Brigade
Brigadier General James L. Kemper
1st Virginia
7th Virginia
11th Virginia
17th Virginia
24th Virginia

Artillery
Wise Virginia Battery

Toombs' Division (temporary)*
Brigadier General Robert Toombs

Toombs' Brigade
Colonel Henry L. Benning
2nd Georgia
15th Georgia
17th Georgia
20th Georgia

Drayton's Brigade
Brigadier General Thomas F. Drayton
50th Georgia
51st Georgia
15th South Carolina
3rd South Carolina Bn.
Phillip's Legion

Anderson's Brigade
Brigadier General George T. Anderson
1st Georgia Regulars
7th Georgia
8th Georgia
9th Georgia
11th Georgia

*This temporary division, split from Anderson's, was created at the onset of the campaign at Leesburg, Virginia.

Walker's Division
Brigadier General John G. Walker

Walker's Brigade
Colonel Van H. Manning
3rd Arkansas
27th North Carolina
46th North Carolina
48th North Carolina
30th Virginia
French's Virginia Battery

Ransom's Brigade
Brigadier General Robert Ransom Jr.
24th North Carolina
25th North Carolina
35th North Carolina
49th North Carolina
Branch's Virginia Battery

Hood's Division
Brigadier General John B. Hood

Hood's Brigade
Colonel William T. Wofford
18th Georgia
Hampton Legion
1st Texas
4th Texas
5th Texas

Law's Brigade
Colonel Evander M. Law
4th Alabama
2nd Mississippi
11th Mississippi
6th North Carolina

Artillery Battalion
German South Carolina Battery
Palmetto South Carolina Battery
Rowan North Carolina Battery

Evans's Brigade
Colonel P. F. Stevens
17th South Carolina
18th South Carolina
22nd South Carolina
23rd South Carolina
Holcombe Legion
Macbeth South Carolina Battery

Corps Artillery

Lee's Battalion
Ashland Virginia Battery
Bedford Virginia Battery
Brooks South Carolina Battery
Eubanks' Virginia Battery
Madison Louisiana Battery
Parker's Virginia Battery

Washington Artillery Battalion
1st Company
2nd Company
3rd Company
4th Company

Jackson's Command
Major General Thomas J. Jackson

Ewell's Division
Brigadier General Alexander R. Lawton

Lawton's Brigade

Colonel Marcellus Douglass

13th Georgia
26th Georgia
31st Georgia
38th Georgia
60th Georgia
61st Georgia

Early's Brigade
Brigadier General Jubal A. Early

13th Virginia
25th Virginia
31st Virginia
44th Virginia
49th Virginia
52nd Virginia
58th Virginia

Trimble's Brigade

Colonel James A. Walker

15th Alabama
12th Georgia
21st Georgia
21st North Carolina

Hay's Brigade
Brigadier General Harry T. Hays
5th Louisiana
6th Louisiana
7th Louisiana
8th Louisiana
14th Louisiana

Artillery
Johnson's Virginia Battery
Louisiana Guard Artillery

A. P. Hill's Light Division
Major General Ambrose P. Hill

Branch's Brigade
Brigadier General Lawrence O. Branch

7th North Carolina
18th North Carolina
28th North Carolina
33rd North Carolina
37th North Carolina

Archer's Brigade
Brigadier General James J. Archer
5th Alabama Battalion
19th Georgia
1st Tennessee Provisional Army
7th Tennessee
14th Tennessee

Artillery Battalion
Crenshaw's Virginia Battery
Fredericksburg Virginia Battery
Pee Dee South Carolina Battery
Purcell Virginia Battery

Gregg's Brigade
Brigadier General Maxcy Gregg
1st South Carolina Provisional Army
1st South Carolina Rifles
12th South Carolina
13th South Carolina
14th South Carolina

Pender's Brigade
Brigadier General William D. Pender
16th North Carolina
22nd North Carolina
34th North Carolina
38th North Carolina

Field's Brigade
Colonel John M. Brockenbrough

40th Virginia
47th Virginia
55th Virginia
22nd Virginia Bn.

Jones' Division
Brigadier General John R. Jones

Stonewall Brigade
Colonel Arnold J. Grigsby
4th Virginia
5th Virginia
27th Virginia
33rd Virginia

Stark's Brigade
Brigadier General William E. Starke
1st Louisiana
2nd Louisiana
9th Louisiana
10th Louisiana
15th Louisiana
Coppens Battalion

Taliaferro's Brigade
Colonel James W. Jackson
47th Alabama
48th Alabama
23rd Virginia
37th Virginia

Andrew's Battalion

Alleghany Virginia Battery

Brockenbrough's Maryland Battery
Danville Virginia Battery
Lee Virginia Battery
Rockbridge Virginia Battery

Jones's Brigade
Cpt. John E. Penn
21st Virginia
42nd Virginia
48th Virginia
1st Virginia Bn.

D. H. Hill's Division
Major General Daniel H. Hill

Ripley's Brigade
Brigadier General Roswell S. Ripley
4th Georgia
44th Georgia
1st North Carolina
3rd North Carolina

Garland's Brigade
Colonel Duncan K. McRae
5th North Carolina
12th North Carolina
13th North Carolina
20th North Carolina
23rd North Carolina

Anderson's Brigade
Brigadier General George B. Anderson
2nd North Carolina
4th North Carolina
14th North Carolina
30th North Carolina

Rodes' Brigade
Brigadier General Robert E. Rodes
3rd Alabama
5th Alabama
6th Alabama
12th Alabama
26th Alabama

Colquitt's Brigade
Colonel Alfred H. Colquitt
13th Alabama
6th Georgia
23rd Georgia
27th Georgia
28th Georgia

Artillery Battalion
Hardaway's Alabama Battery
Jefferson Davis Alabama Battery
Jones' Virginia Battery
King William Virginia Battery

Artillery Reserve
Brigadier General William N. Pendleton

Cutts's Artillery Battalion
Blackshear's Georgia Battery
Irwin's Georgia Battery
Patterson's Georgia Battery
Ross' Georgia Battery

Jones' Artillery Battalion
Morris Virginia Battery
Orange Virginia Battery
Turner's Virginia Battery
Wimbish's Virginia Battery

Miscellaneous Artillery
Cutshaw's Virginia Battery
Dixie Virginia Battery
Magruder Virginia Battery
Rice's Virginia Battery

Cavalry
Major General J.E.B. Stuart

Hampton's Brigade
Brigadier General Wade Hampton
1st North Carolina Cavalry
2nd South Carolina Cavalry
Cobb's Georgia Legion
Jeff Davis Legion

Fitzhugh Lee's Brigade
Brigadier General Fitzhugh Lee
1st Virginia Cavalry
3rd Virginia Cavalry
4th Virginia Cavalry
5th Virginia Cavalry
9th Virginia Cavalry

Robertson's Brigade
Colonel Thomas T. Munford
2nd Virginia Cavalry
7th Virginia Cavalry
12th Virginia Cavalry

Horse Artillery
Chew's Virginia Battery
Hart's South Carolina Battery
Pelham's Virginia Battery

Bibliography

A Committee of the Regimental Association. *History of the Thirty-Fifth Regiment Massachusetts Volunteers, 1862-1865.* Boston: Mills, Knight, & Co., Printers, 1884.

Albert, Allen D. ed. *History of the Forty-Fifth Regiment Pennsylvania Veteran Volunteer Infantry 1861-1865.* Williamsport: Grit Publishing Company, 1912.

Armstrong Jr. Marion V. *Opposing the Second Corps at Antietam: The Fight for the Confederates Left & Center on America's Bloodiest Day.* Tuscaloosa: The University of Alabama Press, 2016.

Armstrong Jr. Marion V. *Unfurl Those Colors! McClellan, Sumner, & the Second Army Corps in the Antietam Campaign.* Tuscaloosa: The University of Alabama Press, 2008.

Bate, Samuel P. *History of Pennsylvania Volunteers, 1861-5; Prepared in Compliance With Acts of the Legislature. Vol. 1.* Harrisburg: B. Singerly, State Printer, 1869.

Brown, Edmund Randolph. *The Twenty-Seventh Indiana Volunteer Infantry in the War of the Rebellion, 1861 to 1865, First Division 12th and 20th Corps.* Monticello, 1899.

Bryant, Edwin E. *History of the Third Regiment of Wisconsin Veteran Volunteer Infantry 1861-1865.* Cleveland: The Arthur H. Clark Company, 1891.

Carman, Ezra A. *The Maryland Campaign of September 1862: Vol. II: Antietam.* Edited by Thomas G. Clemens. El Dorado Hills: Savas Beatie LLC, 2012.

Carman, Ezra A. and Emmor B. Cope. "Atlas of the Battlefield of Antietam, prepared under the direction of the Antietam Battlefield Board, Lieut. Col. Geo. W. Davis, U.S.A., president, Gen. E.A. Carman, U.S.V., Gen. H. Heth, C.S.A. Surveyed by Lieut. Col. E. B. Cope, engineer, H.W. Mattern, assistant engineer, of the Gettysburg National Park. Drawn by Charles H. Ourand, 1899. Position of troops by Gen. E. A. Carman. Published by authority of the Secretary of War, under the direction of the Chief of Engineers, U.S. Army, 1908.", 1904, Revised Edition 1908, Library of Congress.

Chapin, L. N. *A Brief History of the Thirty-Fourth Regiment N. Y. S. V. Embracing a Complete Roster of All Officers and Men and A Full Account of the Dedication of the Monument on the Battlefield of Antietam September 17, 1862 With Numerous Illustrations.* New York, 1903.

Clark, Walter. *Histories of the Several Regiments and Battalions from North Carolina, in the Great War 1861-'65. Written By Members of Their Respective Commands. Vol. II.* Goldsboro: Book and Job Printers, 1901.

Croffut, W. A. and John M. Morris. *The Military and Civil History of Connecticut During the War of 1861-65: Comprising a Detailed Account of the Various Regiments and Batteries, Through March, Encampment, Bivouac, and Battle: Also Instances of Distinguished Personal Gallantry, and Biographical Sketches of Many Heroic Soldiers: Together With Record of the Patriotic Action of Citizens at Home, and of the Liberal Support Furnished by the State in its Executive and Legislative Departments.* Boston: Geo. C. Rand & Avery Stereotypers and Printers, 1868.

Crowell, Joseph E. *The Young Volunteer; The Everyday Experiences of a Soldier Boy in the Civil War.* Paterson: The Call, 1906.

Cunningham, D. and W. W. Miller. *Antietam: Report of the Ohio Antietam Battlefield Commission.* Springfield: Springfield Publishing Company, State Printer, 1904.

Dawes, Rufus R. *Service with the Sixth Wisconsin Volunteers.* Marietta: E. R. Alderman & Sons, 1890.

Dickert, Augustus D. *History of Kershaw's Brigade with Complete Roll of Companies, Biographical Sketches, Incidents, Anecdotes, etc.* Newberry: Elbert H. Aull Company, 1899.

Eddy, Richard. *History of the Sixtieth Regiment New York State Volunteers: From the Commencement of its Organization in July, 1861, to its Public Reception at Ogdensburgh as a Veteran Command, January 7th, 1864.* Philadelphia: Published by the Author, 1864.

Frassanito, William A. *Antietam: The Photographic Legacy of America's Bloodiest Day.* Gettysburg: Thomas Publications, 1978.

Gottfried, Bradley M. *The Maps of Antietam: An Atlas of the Antietam (Sharpsburg) Campaign, Including the Battle of South Mountain, September 2-20, 1862.* El Dorado Hills: Savas Beatie LLC, 2012.

Gould, John M. *History of the First-Tenth-Twenty-ninth Maine Regiment. In Service Of The United States From May 3rd, 1861, to June 21st, 1866.* Portland: Stephen Berry, 1871.

Graham, Mathew J. *The Ninth Regiment, New York Volunteers (Hawkins' Zouaves).* New York: E. P. Coby & Co. Printers, 1900.

Hinkley, Julian Wisner. *A Narrative of Service With the Third Wisconsin Infantry.* Democrat Printing Co., State Printer, 1912.

Hood, J. B. *Advance and Retreat: Personal Experiences in the United States and Confederate Armies.* Philadelphia: Burk & M'Fetridge, 1880.

Horton & Teverbaugh, *A History of the Eleventh regiment, (Ohio Volunteer Infantry,) Containing the Military Record ... of Each Officer and Enlisted Man of the Command-a List of Deaths-An Account of the Veterans-Incidents of the Field and Camp-names of the Three Months' Volunteers, etc., etc..* Dayton: W. J. Shuey, Printer and Publisher, "Telescope Office", 1866.

Hyde, Thomas W. *Following the Greek Cross or, Memories of the Sixth Army Corps.* Cambridge: The Riverside Press, 1894.

Jackson, Lyman. *History of the Sixth New Hampshire regiment in the War for the Union.* Concord: Republican Press Association, Railroad Square, 1891.

Johnson, Curt and Richard C. Anderson Jr. *Artillery Hell: The Employment of Artillery at Antietam.* College Station: Texas A&M University Press, 1995.

Lord, A. M., Edward O. *History of the Ninth Regiment New Hampshire Volunteers in the War of the Rebellion.* Concord: Republican Press Association, 1895.

Parker, Thomas H., *History of the 51st regiment of P.V. and V.V., From its Organization, at Camp Curtin, Harrisburg, Pa., in 1861, to its Being Mustered Out of the United States Service at Alexandria, Va., July 27th, 1865.* Philadelphia: King & Baird, Printers, 1869.

Page, Charles, D. *History of the Fourteenth Regiment, Connecticut Vol. Infantry*. Meridan: The Boston Printing Co. 1906.

Pickerill, W. N. ed. *Indiana at Antietam: Report of the Indiana Antietam Monument Commission and Ceremonies at the Dedication of the Monument*. Indianapolis: The Aetna Press, 1911.

Quint, Alonzo H. *The Record of the Second Massachusetts Infantry, 1861-65*. Boston: James P. Walker, 1867.

Rawle, William Brooke. *History of the Third Pennsylvania Cavalry, Sixtieth Regiment Pennsylvania Volunteers, in the American Civil War 1861-1865*. Philadelphia: Franklin Printing Company, 1905.

Regimental Committee, The. *History of the One Hundredth and Twenty-Fifth Regiment Pennsylvania Volunteers 1862-1863*. Philadelphia: J. B. Lippincott Company, 1906.

Seville, William P. *History of the First regiment, Delaware Volunteers, from the Commencement of the "Three Months' Service" to the Final Muster-out at the Close of the Rebellion*. Wilmington: The Historical Society of Delaware, 1884.

Thomson, O. R. Howard and William H. Rauch. *History of the "Bucktails" Kane Rifle Regiment of the Pennsylvania Reserve Corps (13th Pennsylvania Reserves, 42nd of the Line)*. Philadelphia: Electric Printing Company, 1906.

Tombs, Samuel. *Reminiscences of the War, Comprising a Detailed Account of the Experiences of the Thirteenth Regiment New Jersey Volunteers in Camp, on the March, and in Battle*. Orange: Journal Office, 1878.

United States House of Representatives. *Report of the Joint Committee on the Conduct of the War: In Three Parts*. Washington D. C.: Government Printing Office, 1863.

U. S. War Department. *The War of the Rebellion: A Compilation of the Official Records of the Union and Confederate Armies*. 128 vols. Washington D. C.: Government Printing Office, 1880-1901.
Vol. 2 Bibliography.

Washburn, George H. *A Complete Military History and Record of the 108th Regiment N.Y. Vols., from 1862 to 1894. Together with Roster, Letters, Rebel Oaths of Allegiance, Rebel Passes, Reminiscences, Life Sketches, Photographs, etc., etc.* Rochester: Press of E. R. Andrews, 1894.

Notes

Prelude to Battle

[1] Regimental History Committee, *History of the Third Pennsylvania Cavalry, Sixtieth Regiment Pennsylvania Volunteers, In the American Civil War 1861-1865*, (Philadelphia: Franklin Printing Company, 1905), 120-22.
[2] Ibid, 122.
[3] O. R. Howard Thomson and William H. Rauch, *History of the "Bucktails," Kane Rifle Regiment of the Pennsylvania Reserve Corps, (13th Pennsylvania Reserves, 42nd of the Line)*, (Philadelphia: Electric Printing Company, 1906), 209-211.
[4] U. S. War Department, *The War of the Rebellion: A Compilation of the Official Records of the Union and Confederate Armies*, (Washington: Government Printing Office, 1887), Series I, vol. 51, part I, 148. Hereafter the work will be cited as OR.
[5] Thomson and Rauch, *History of the "Bucktails*," 211.
[6] *OR*, vol. 19, part I, 923.

The First Corps Begins the Battle

[1] Carman, Ezra A. *The Maryland Campaign of September 1862: Vol. II: Antietam*, ed. Thomas G. Clemens, (El Dorado Hills: Savas Beatie LLC, 2012), 41-42.
[2] Carman, *The Maryland Campaign*, 55.
[3] Ibid, 56.
[4] Carman, *The Maryland Campaign*, 150-52.
[5] Ibid, 112.
[6] Ibid, 59.
[7] Ibid, 61-62.
[8] *OR*, vol. 19, part I, 977. All subsequent references are to Series I, vol. 19, part I unless otherwise indicated.
[9] *OR*, 1033.
[10] Carman, *The Maryland Campaign*, 82-83.
[11] Ibid, 76-77
[12] Ibid, 76.
[13] *OR*, 1017.
[14] *OR*, 233, 1017; Dawes, Rufus R., *Service with the Sixth Wisconsin Volunteers*, (Marietta: E. R. Alderman & Sons, 1890), 90-91.
[15] Hood, J. B., *Advance and Retreat: Personal Experiences in the United States and Confederate Armies*, (Philadelphia: Burk & M'Fetridge, 1880), 42-43; Carman, *The Maryland Campaign*, 88-89.
[16] OR, 265-66, 937-38.
[17] Ibid, 243-44, 251.
[18] *OR*, 977; Carman, *The Maryland Campaign*, 91.
[19] *OR*, 934-36.
[20] *OR*, 270; Carman, *The Maryland Campaign*, 94.
[21] *OR*, 229.
[22] *OR*, 266, 937-38; Carman, *The Maryland Campaign*, 90-91.
[23] *OR*, vol. 51, part I, 139.
[24] *OR*, 936; Carman, *The Maryland Campaign*, 94.
[25] Carman, *The Maryland Campaign*, 91.
[26] *OR*, 274; Carman, *The Maryland Campaign*, 96-96.
[27] *OR*, 488; Carman, *The Maryland Campaign*, 117; Gould, John M. *History of the First-Tenth-Twenty-ninth Maine Regiment. In Service Of The United States From May 3rd, 1861, to June 21st, 1866*, (Portland: Stephen Berry, 1871), 236.

The Twelfth Corps Attacks

[1] *OR*, 932; Carman, *The Maryland Campaign*, 102.
[2] *OR*, 229, 243-44. 251, 930, 935; Carman, *The Maryland Campaign*, 103-06.
[3] *OR*, 938.

[4] OR, 1017; Carman, *The Maryland Campaign*, 107, 119-20.
[5] OR, 487, 493-94.
[6] Carman, *The Maryland Campaign*, 120.
[7] OR, 493.
[8] Carman, *The Maryland Campaign*, 166-67. The exact circumstances of Mansfield's death are shrouded in controversy. According to Carman, there are at least seven different accounts. The version told by the 10th Maine's John Gould is generally taken to be the most accurate, and the basis for the current location of Mansfield's wounding monument at the park.
[9] OR, 498-99; Carman, *The Maryland Campaign*, 125-26. Brown, Edmund Randolph, *The Twenty-Seventh Indiana Volunteer Infantry in the War of the Rebellion, 1861 to 1865, First Division 12th and 20th Corps*, (Monticello, 1899), 243-250.
[10] OR, 490-91; Carman, *The Maryland Campaign*, 157-58.
[11] Carman, *The Maryland Campaign*, 136-37. Cunningham, D. and W. W. Miller, *Antietam: Report of the Ohio Antietam Battlefield Commission*, (Springfield: Springfield Publishing Company, State Printer, 1904), 47.
[12] OR, 488.
[13] OR, 1,043-45; Carman, *The Maryland Campaign*, 132-33.
[14] OR, 507, 508; Carman, *The Maryland Campaign*, 137-38.
[15] Carman, *The Maryland Campaign*, 158-59.
[16] Eddy, Richard, *History of the Sixtieth Regiment New York State Volunteers: From the Commencement of its Organization in July, 1861, to its Public Reception at Ogdensburgh as a Veteran Command, January 7th, 1864*, (Philadelphia: Published by the Author, 1864), 180-81.
[17] OR, 490-91; Carman, *The Maryland Campaign*, 157-58.
[18] OR, 228; Carman, *The Maryland Campaign*, 161, 176-77.
[19] OR, 970-71.
[20] OR, 219; United States House of Representatives, *Report of the Joint Committee on the Conduct of the War: In Three Parts*, (Washington D. C.: Government Printing Office, 1863), 581.

The West Woods

[1] OR, 492; Carman, *The Maryland Campaign*, 179.
[2] Armstrong Jr. Marion V., *Unfurl Those Colors! McClellan, Sumner, & the Second Army Corps in the Antietam Campaign*, (Tuscaloosa: The University of Alabama Press, 2008), 174-79.
[3] OR, 970; Carman, *The Maryland Campaign*, 180.
[4] Regimental Committee, The, *History of the One Hundredth and Twenty-Fifth Regiment Pennsylvania Volunteers 1862-1863*, (Philadelphia: J. B. Lippincott Company, 1906), 72.
[5] OR, 909; Carman, *The Maryland Campaign*, 181.
[6] OR, 315-16.
[7] OR, 909; Carman, *The Maryland Campaign*, 181.
[8] OR, 865, 971; Carman, *The Maryland Campaign*, 188-89.
[9] OR, 228; Carman, *The Maryland Campaign*, 175.
[10] OR, 308; Carman, *The Maryland Campaign*, 177.
[11] OR, 313.
[12] OR, 883; Carman, *The Maryland Campaign*, 200-201; Regimental Committee, *History of the One Hundredth and Twenty-Fifth Regiment*, 72-73.
[13] OR, 860, 874, 877.
[14] OR, 306, 316, 321; Carman, *The Maryland Campaign*, 201-204; The 34th New York reported that Sedgwick ordered them out of the woods. Howard likewise asserts that Sumner moved the 72nd Pennsylvania.
[15] OR, 320.
[16] Carman, *The Maryland Campaign*, 208-09. Many in Howard's brigade dispute that they were routed, excepting the 72nd Pennsylvania.
[17] OR, 868; Carman, *The Maryland Campaign*, 206, 234-35.
[18] Carman, *The Maryland Campaign*, 202.
[19] OR, 313; Carman, *The Maryland Campaign*, 210-13.
[20] OR, 309, 868-69, 883; Carman, *The Maryland Campaign*, 213-14.
[21] OR, 865; Carman, *The Maryland Campaign*, 236; Armstrong Jr. Marion V., *Opposing the Second Corps at Antietam: The Fight for the Confederates Left & Center on America's Bloodiest Day*, (Tuscaloosa: The University of Alabama Press, 2016), 55-56.
[22] Carman, *The Maryland Campaign*, 210-11.

23 *OR*, 311, 314; Carman, *The Maryland Campaign*, 211-12.
24 *OR*, 152, 236.
25 *OR*, 317, 323; Carman, *The Maryland Campaign*, 219.
26 *OR*, 219.
27 *OR*, 865; Carman, *The Maryland Campaign*, 236; Armstrong, *Opposing the Second Corps*, 55-56.
28 Armstrong, *Opposing the Second Corps*, 55-56.
29 *OR*, 865; Armstrong, *Opposing the Second Corps*, 58.
30 *OR*, 323-24; Carman, *The Maryland Campaign*, 236. Armstrong, *Opposing the Second Corps*, 58. Based on a lack of primary evidence, there is some doubt whether the 8th South Carolina advanced far enough to fire into the flank of the 1st Delaware. However, General French in his report clearly stated that Confederates, "which had succeeded in breaking the center division [Greene's] of the line of battle," fired upon Weber's right flank. The 8th is still the most likely candidate.
31 *OR*, 506-09.
32 Armstrong, *Opposing the Second Corps*, 62.
33 *OR*, 308, 915; Armstrong, *Opposing the Second Corps*, 59.
34 Armstrong, *Opposing the Second Corps*, 62.
35 Ibid., 69.
36 *OR*, 872, 1,037.
37 Armstrong, *Opposing the Second Corps*, 59-61.
38 Ibid., 61-62.
39 *OR*, 495.
40 *OR*, 502; Tombs, Samuel, *Reminiscences of the War, Comprising a Detailed Account of the Experiences of the Thirteenth Regiment New Jersey Volunteers in Camp, on the March, and in Battle*, (Orange: Journal Office, 1878), 20.
41 *OR*, 505.
42 Ibid, 505; Carman, *The Maryland Campaign*, 306-08.
43 *OR*, 505, 507-08.
44 Ibid., 515.
45 Carman, *The Maryland Campaign*, 315.
46 Ibid., 310.
47 *OR*, 872.
48 Carman, *The Maryland Campaign*, 316.
49 *OR*, 515.
50 *OR*, 325; Carman, *The Maryland Campaign*, 303-04.
51 *OR*, 915; Carman, *The Maryland Campaign*, 318-19; Clark, Walter. *Histories of the Several Regiments and Battalions from North Carolina, in the Great War 1861-'65. Written By Members of Their Respective Commands. Vol. II.* (Goldsboro: Book and Job Printers, 1901), 434.
52 Carman, *The Maryland Campaign*, 317-18.

The Sunken Road

1 Armstrong, *Unfurl Those Colors!*, 178.
2 *OR*, 323-24; Carman, *The Maryland Campaign*, 236. Armstrong, *Opposing the Second Corps*, 58.
3 Washburn, George H., *A Complete Military History and Record of the 108th Regiment N.Y. Vols., from 1862 to 1894. Together with Roster, Letters, Rebel Oaths of Allegiance, Rebel Passes, Reminiscences, Life Sketches, Photographs, etc., etc.*, (Rochester: Press of E. R. Andrews, 1894), 24-25.
4 *OR*, 332, 337.
5 *OR*, 872, 1,037.
6 Ibid.
7 Seville, William P., *History of the First regiment, Delaware Volunteers, from the Commencement of the "Three Months' Service" to the Final Muster-out at the Close of the Rebellion.* (Wilmington: The Historical Society of Delaware, 1884), 48-49.
8 *OR*, 1,051; Carman, *The Maryland Campaign*, 260.
9 Carman, *The Maryland Campaign*, 270; Armstrong, *Opposing the Second Corps*, 104.
10 *OR*, 327; Carman, *The Maryland Campaign*, 262; Armstrong, *Opposing the Second Corps*, 105-07.
11 Armstrong, *Opposing the Second Corps*, 107.
12 Carman, *The Maryland Campaign*, 262-63.
13 Ibid., 271.
14 Ibid., 279-80; Armstrong, *Unfurl Those Colors!*, 228.
15 *OR*, 1,037-38; Carman, *The Maryland Campaign*, 278-79.

[16] *OR*, 289; Carman, *The Maryland Campaign*, 282.
[17] *OR*, 331, 332, 334-35.
[18] *OR*, 289; Carman, *The Maryland Campaign*, 282.
[19] Carman, *The Maryland Campaign*, 279-80.
[20] *OR*, 290.
[21] Ibid., 872.
[22] Carman, *The Maryland Campaign*, 289-90; Armstrong, *Opposing the Second Corps*, 120.
[23] *OR*, 285-86, 292, 299; Carman, *The Maryland Campaign*, 289-90; Armstrong, *Opposing the Second Corps*, 120-22; Armstrong, *Unfurl Those Colors!*, 232-239.
[24] *OR*, 327, 329, 330; Carman, *The Maryland Campaign*, 286.
[25] *OR*, 290, 299, 304.
[26] Ibid., 327, 329, 330; *Opposing the Second Corps*, 128-29.
[27] *OR*, 304.
[28] Ibid., 872.
[29] *OR*, 411; Carman, *The Maryland Campaign*, 287.
[30] *OR*, 402, 409.
[31] Ibid., 302, 303.
[32] Ibid., 286; Armstrong, *Unfurl Those Colors!*, 243-45.
[33] *OR*, 343-44.
[34] Ibid., 820, 956-57.
[35] Carman, *The Maryland Campaign*, 337-38.
[36] Ibid.
[37] *OR*, 412; Hyde, Thomas W., *Following the Greek Cross or, Memories of the Sixth Army Corps*, (Cambridge: The Riverside Press, 1894), 99-100.
[38] *OR*, 412; Carman, *The Maryland Campaign*, 346-47; Hyde, *Following the Greek Cross*, 100-101.
[39] Carman, *The Maryland Campaign*, 347-48; Hyde, *Following the Greek Cross*, 101-102.
[40] *OR*, 412-43; Carman, *The Maryland Campaign*, 347-48; Hyde, *Following the Greek Cross*, 101-103.

The Middle Bridge

[1] Carman, *The Maryland Campaign*, 359.
[2] Ibid.
[3] Ibid.
[4] *OR*, 358.
[5] Ibid., 360; Carman, *The Maryland Campaign*, 368.
[6] *OR*, 357, 359, 36; Carman, *The Maryland Campaign*, 379-80.
[7] *OR*, 939.
[8] *OR*, 1,044-45; Carman, *The Maryland Campaign*, 380.
[9] *OR*, 351, 363; Carman, *The Maryland Campaign*, 384-85.
[10] *OR*, 1,031, 1.025; Carman, *The Maryland Campaign*, 384-85.
[11] *OR*, 441; Carman, *The Maryland Campaign*, 384-85.

Burnside's Bridge

[1] *OR*, 450-51; Carman, *The Maryland Campaign*, 402.
[2] *OR*, 419; Croffut, W. A. and John M. Morris. *The Military and Civil History of Connecticut During the War of 1861-65: Comprising a Detailed Account of the Various Regiments and Batteries, Through March, Encampment, Bivouac, and Battle: Also Instances of Distinguished Personal Gallantry, and Biographical Sketches of Many Heroic Soldiers: Together With Record of the Patriotic Action of Citizens at Home, and of the Liberal Support Furnished by the State in its Executive and Legislative Departments*, (Boston: Geo. C. Rand & Avery Stereotypers and Printers, 1868), 266.
[3] Croffut and Morris, *Military and Civil History*, 266-67.
[4] *OR*, 473; Carman, *The Maryland Campaign*, 412; Horton & Teverbaugh, *A History of the Eleventh regiment, (Ohio Volunteer Infantry,) Containing the Military Record ... of Each Officer and Enlisted Man of the Command-a List of Deaths-An Account of the Veterans-Incidents of the Field and Camp-names of the Three Months' Volunteers, etc., etc.*, (Dayton: W. J. Shuey, Printer and Publisher, "Telescope Office", 1866), 74-75.
[5] *OR*, 472-73; Carman, *The Maryland Campaign*, 412.

[6] Carman, *The Maryland Campaign*, 415.
[7] *OR*, 444.
[8] *OR*, 447; Carman, *The Maryland Campaign*, 415-16; Jackson, Lyman, *History of the Sixth New Hampshire regiment in the War for the Union*, (Concord: Republican Press Association, Railroad Square, 1891), 104; Lord, A. M., Edward O. *History of the Ninth Regiment New Hampshire Volunteers in the War of the Rebellion*, (Concord: Republican Press Association, 1895), 105-06.
[9] *OR*, 444.
[10] Carman, *The Maryland Campaign*, 417-19; Parker, Thomas H., *History of the 51st regiment of P.V. and V.V., From its Organization, at Camp Curtin, Harrisburg, Pa., in 1861, to its Being Mustered Out of the United States Service at Alexandria, Va., July 27th, 1865*, (Philadelphia: King & Baird, Printers, 1869), 232-33.
[11] A Committee of the Regimental Association. *History of the Thirty-Fifth Regiment Massachusetts Volunteers, 1862-1865*, (Boston: Mills, Knight, & Co., Printers, 1884), 40-41.
[12] Carman, *The Maryland Campaign*, 419-20; Parker, *History of the 51st regiment of P.V*, 233.
[13] *OR*, 472.
[14] *OR*, 452-53; Carman, *The Maryland Campaign*, 425-28.
[15] *OR*, 451; Graham, Mathew J. *The Ninth Regiment, New York Volunteers (Hawkins' Zouaves)*, (New York: E. P. Coby & Co. Printers, 1900), 289-90.
[16] *OR*, 456.
[17] Committee, *Thirty-Fifth Regiment Massachusetts*, 42.

Ninth Corp Advance and Hill's Counterattack

[1] *OR*, 451.
[2] Carman, *The Maryland Campaign*, 451.
[3] *OR*, 438-39; Carman, *The Maryland Campaign*, 444-46.
[4] *OR*, 441; Carman, *The Maryland Campaign*, 443-44; Bate, Samuel P., *History of Pennsylvania Volunteers, 1861-5; Prepared in Compliance With Acts of the Legislature, Vol. 1*, (Harrisburg: B. Singerly, State Printer, 1869), 1,061.
[5] *OR*, 453.
[6] *OR*, 441; Carman, *The Maryland Campaign*, 443-44.
[7] *OR*, 905; Carman, *The Maryland Campaign*, 452-56; Graham, *The Ninth Regiment*, 295-96, 317-19.
[8] *OR*, 453.
[9] *OR*, 988.
[10] Ibid.
[11] Carman, *The Maryland Campaign*, 443.
[12] Carman, *The Maryland Campaign*, 458-59; Croffut and Morris, *Military and Civil History*, 272.
[13] Carman, *The Maryland Campaign*, 459.
[14] Ibid.
[15] *OR*, 891; vol. 51, part I, 164.
[16] Carman, *The Maryland Campaign*, 462. Rodman was moving between Fairchild's brigade and the 8th Connecticut when he was shot. Colonel Fairchild saw his rider less horse cross from west to east the stone wall his brigade had captured.
[17] *OR*, 988. 996-97.
[18] Ibid., 993-94.
[19] Ibid.
[20] Ibid. 454.
[21] Ibid.
[22] *OR*, 466, 467-68; Carman, *The Maryland Campaign*, 468-70.
[23] *OR*, 435; Carman, *The Maryland Campaign*, 484.
[24] *OR*, 1,000-01; Carman, *The Maryland Campaign*, 472-73.
[25] *OR*, 468.
[26] Ibid. 470, 1,001.
[27] Ibid. 470, 997.
[28] Carman, *The Maryland Campaign*, 476.
[29] Carman, *The Maryland Campaign*, 476; Committee, *Thirty-Fifth Regiment Massachusetts*, 46-47.
[30] *OR*, 1,001; Carman, *The Maryland Campaign*, 479; Clark, Walter. *Histories of the Several Regiments and Battalions from North Carolina, in the Great War 1861-'65. Written By Members of Their Respective Commands. Vol. II*, (Goldsboro: Book and Job Printers, 1901), 32-33.
[31] Committee, *Thirty-Fifth Regiment Massachusetts*, 47.

About the Author

Brad Butkovich has a Bachelor of Arts degree in history from Georgia Southern University. He has published several books on the American Civil War including studies on the Battle of Pickett's Mill, Allatoona Pass, and the *Visual Antietam* series. He has always had a keen interest in Civil War history, photography and cartography, all of which have come together in his current projects.

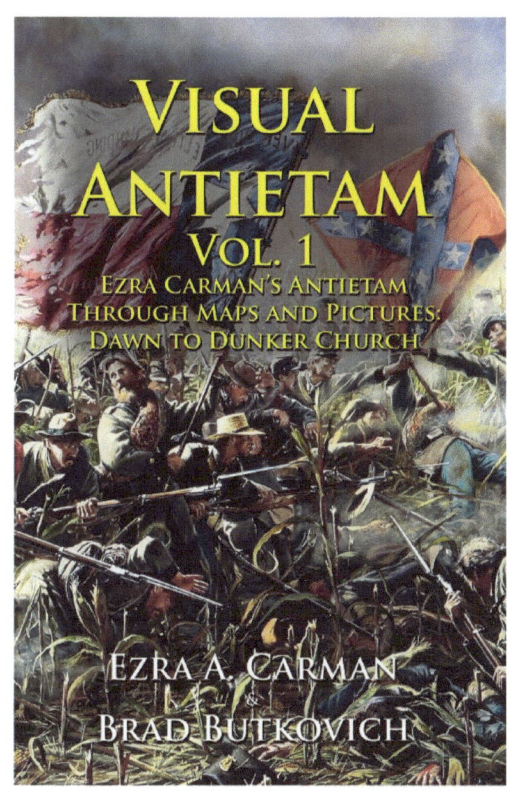
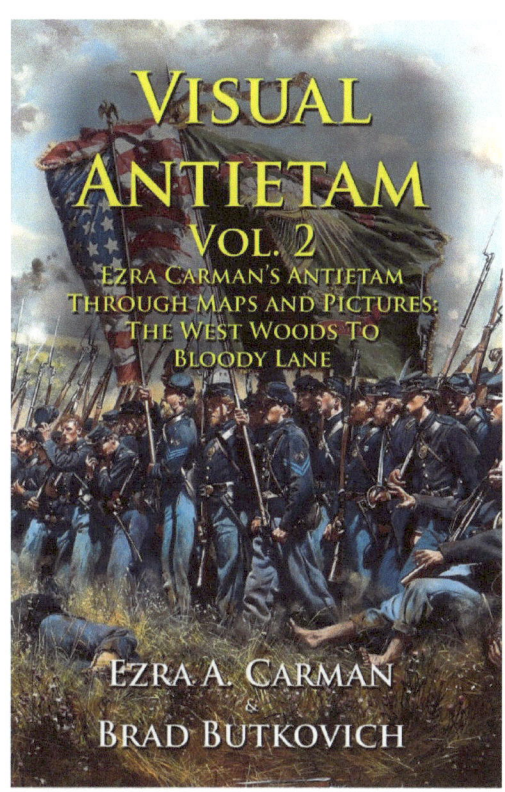
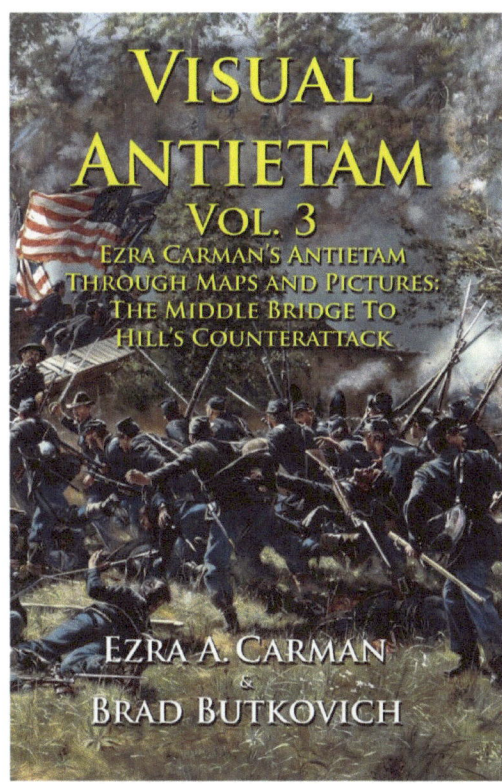